T3-BWZ-753

Pasture and rangeland provide 80-90 percent of the feed to nourish the nation's beef herds and sheep flocks.

DON W. GRAFFIS, Ph.D.

Extension Agronomist
University of Illinois at Urbana – Champaign

E. M. JUERGENSON, Ph.D.

Teacher Educator (Retired)
University of California, Davis

MALCOLM H. McVICKAR, Ph.D.

Late Chief of Agronomic Education
National Plant Food Institute
Washington, D.C.

Approved Practices in Pasture Management

THE INTERSTATE
PRINTERS & PUBLISHERS, INC.

Danville, Illinois

 Prior editions: 1956, 1963, 1974. Printed in the United States of America.

Library of Congress Catalog Card No. 84-80673

1 2 3
4 5 6
7 8 9

ISBN 0-8134-2449-6

Preface

Forage production has always played an important role in American agriculture. With a continually expanding world population and higher standards for human nutrition, the demand for animal proteins—meat, milk, and eggs—is increasing. Forage crops play a major role in the economic production of these products. Also, the need to conserve our soil resource makes efficient forage production of greater and greater importance.

In the past, in contrast with producers of feed grains, forage producers have not kept abreast with practices proven profitable by research and experience on farms having highly successful forage-producing programs. In most instances, the less productive, more erosive soils are allotted the forage crops while the feed grains are grown on the most productive soils. More than a billion acres of land in the United States is used for forage production. Much of this area has limited crop production potential due to low rainfall, low natural fertility, or steep slopes, restricting crop choice. The demand for food, the demand for conserving our soil resource, and the demand for economic soundness will place continuing pressure on American agriculture to adopt the highest technological practices known to produce and utilize forage crops. Farmers will need to rapidly adopt the new high-yielding varieties possessing insect and disease resistance. They will need to apply fertilizer to their forage land in many cases. They will need to adjust their stocking rates and choice of animals, and to regulate grazing periods. Efficient forage production and utilization demands a high degree of managerial ability, technical information, and operational skill.

Forage users cannot overlook the inherent productiveness of the animals. Beef, dairy, and swine performance testing stations throughout the United States and abroad are helpful in selecting breeding stock with greater feed efficiency.

The value of good pastures can be measured by the increased animal units carried on a given land area, the improved productivity of the animals, the amount of grain and hay replaced to sustain the herd throughout the year, and the reduction of soil erosion. Properly managed, the acres devoted to pastures and hay can be the most productive on the farm. *Approved Practices in Pasture Management* covers activities associated with the establishment, management, and utilization of pastures. The "whys," "hows," and "whens" of doing the job are presented in condensed form. Specific information has been offered where possible, but where practices vary because of climatic or other conditions, basic information and illustrations are presented which make it possible for the reader to tailor the program best suited to his or her enterprise.

This edition includes a chapter on pasture for hogs and a chapter on pasture for poultry, although it is recognized that because of labor and capital requirements, there is much less use of pasture for swine and poultry today than in earlier years. However, the practices outlined in these two chapters should be valuable to those who are still using pastures in producing swine and small farm flocks.

Approved Practices in Pasture Management should be of special value to teachers of agriculture and their students, FFA and 4-H Club members, county farm advisors and agents, and all others interested in improved forage production. It is the sincere hope of the authors that this book will prove helpful to livestock producers whether they are actively engaged in a livestock enterprise or simply have an interest in livestock production.

Table of Contents

CHAPTER 1

Pastures—America's Number One Crop

More acres are devoted to pastures than to any other crop. In fact, the land used for pastures occupies as many acres in the United States as all other crops combined.

The National Inventory of Soil and Water Conservation Needs supplies some interesting data concerning the agricultural resources of the 50 states making up our nation. Roughly, three-fourths of the total area (over 1½ billion acres) is agricultural land, of which approximately two-thirds is privately owned. Thus privately owned land supplies most of the nation's food, fiber, timber, water, and wildlife, and its owners must assume major responsibility for appropriate management according to land use capabilities.

The following division of private agricultural land of the mainland area is shown by the report:

31% for cropland (447 million acres)
33% for pasture and range (485 million acres)
31% for forest and woodland (450 million areas)
5% for other uses (60 million acres)

> Four-fifths of this land is about equally divided between four land capability classes, half of it suited, and half of it not suited to regular cultivation.

One-fifth is in land capability Class II.
One-fifth is in land capability Class III.
One-fifth is in land capability Class VI.
One-fifth is in land capability Class VII.

When one adds to the 485 million acres listed as used for pasture the acreage of woodland actually used for grazing and the vast

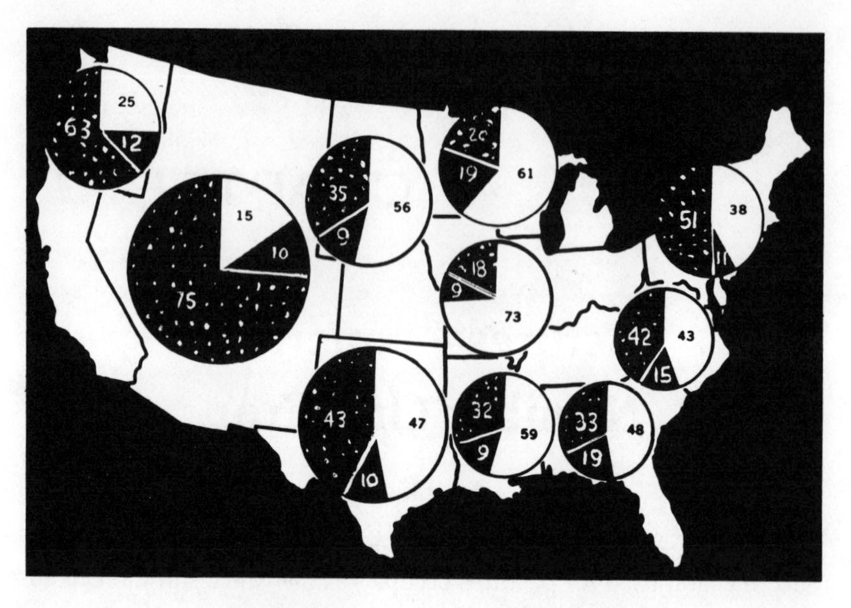

NORTHEAST
□ 39,193,000 acres
■ 10,848,000 acres
▩ 52,576,000 acres

SOUTHEAST
□ 54,165,000 acres
■ 21,298,000 acres
▩ 37,454,000 acres

LAKE STATES
□ 66,992,000 acres
■ 20,579,000 acres
▩ 21,698,000 acres

DELTA STATES
□ 48,163,000 acres
■ 7,491,000 acres
▩ 26,520,000 acres

NORTHERN PLAINS
□ 103,595,000 acres
■ 16,870,000 acres
▩ 63,472,000 acres

SOUTHERN PLAINS
□ 94,198,000 acres
■ 20,066,000 acres
▩ 88,117,000 acres

CORN BELT
□ 111,578,000 acres
■ 14,457,000 acres
▩ 27,238,000 acres

MOUNTAIN
□ 42,652,000 acres
■ 26,779,000 acres
▩ 206,485,000 acres

APPALACHIAN
□ 48,661,000 acres
■ 17,057,000 acres
▩ 47,310,000 acres

PACIFIC
□ 27,897,000 acres
■ 13,258,000 acres
▩ 69,925,000 acres

□ Suitable for regular cultivation (Class I-III)
■ Suitable for occasional cultivation (Class IV)
▩ Not suitable for cultivation (Class V-VIII)

Courtesy, USDA

Fig. 1.1—Land capability by regions.

acreages of state and federally owned lands that are grazed, the aggregate acreage pastured in the 50 states is well over a billion acres.

Although a more detailed description of the eight land capability classes will be given later, brief reference and descriptions of them seem appropriate at this time.

Land Suitable for Regular Cultivation and for Other Uses

Class I soils have no conditions that limit their use and can be safely cultivated without special conservation treatment.

Class II soils have some natural conditions that limit the kinds of plants they can produce or that call for some easily applied conservation practices when they are cultivated.

Class III soils have more serious or more numerous limitations than Class II soils, such as slope, sandiness, shallowness, or too much or too little water. They are more restricted in use for cultivated crops and require application of more conservation measures to be kept producing efficiently.

Land Suitable for Occasional Cultivation and for Other Uses

Class IV soils have severe limitations that restrict the kinds of plants they can grow, and require careful management if cultivated.

Land Generally Not Suitable for Cultivation and for Other Uses

Class V soils have very little erosion hazard but have some condition such as stoniness, droughtiness, or overflow which is impractical to remove and thus limits their use to pasture, range, woodland, etc.

Class VI soils have severe limitations making them generally unsuited for cultivation and restricting their use to pasture, range, woodland, wildlife cover, or water supply.

Class VII soils have very severe limitations that restrict their use to pasture, range, woodland, water supply, recreation, or wildlife food and cover with careful management.

Class VIII soils and land forms have limitations that prevent their use for commercial plant production and restrict their use to water supply, recreation, or wildlife food and cover with careful protection.

The most pressing needs of the pasture and rangeland of our nation are protection from overgrazing, fire, erosion, insects and diseases, brush, and weeds. The overall need is to improve vegetative cover for the land and to manage the land according to its best use capability.

According to the United States Department of Agriculture, there are 10 farm production regions in the United States. See Fig. 1.2 for names of regions and states included in each region.

Assuming no major changes in agricultural policy, locally estimated change in land use should show little difference from the National Inventory of Soil and Water Conservation Needs report per-

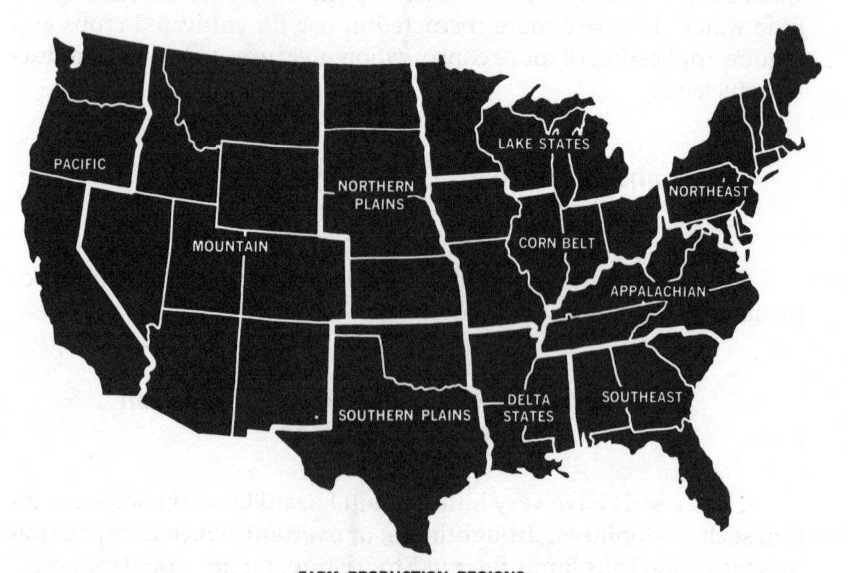

FARM PRODUCTION REGIONS

NORTHEAST: Maine, New Hampshire, Vermont, Massachusetts, Rhode Island, Connecticut, New York, New Jersey, Pennsylvania, Delaware, Maryland • LAKE STATES: Michigan, Wisconsin, Minnesota • CORN BELT: Ohio, Indiana, Illinois, Iowa, Missouri • NORTHERN PLAINS: North Dakota, South Dakota, Nebraska, Kansas • APPALACHIAN: Virginia, West Virginia, North Carolina, Kentucky, Tennessee • SOUTHEAST: South Carolina, Georgia, Florida, Alabama • DELTA STATES: Mississippi, Arkansas, Louisiana • SOUTHERN PLAINS: Oklahoma, Texas • MOUNTAIN: Montana, Idaho, Wyoming, Colorado, New Mexico, Arizona, Utah, Nevada • PACIFIC: Washington, Oregon, California

Courtesy, USDA

Fig. 1.2—Farm production regions.

centages in the acreage devoted to each use. Cropland is expected to decline and pasture and range to increase about 3 percent each. Forest and woodland are expected to decline about 2 percent, at the national level.

It is estimated that cropland will increase in the Pacific and Mountain farm production regions, about hold its own in the Corn Belt, and decline in the seven other regions. It is anticipated that forest and woodland will increase in the Northeast and Southeast farm production regions, about hold their own in the Appalachian region, and decline in the seven other regions. In the pasture and range category, the future will probably see increases in the Northern Plains, Southern Plains, Delta, Mountain, Appalachian, and Southeast farm production regions, with corresponding decreases in the four other regions.

These changes will involve drainage, clearing, irrigation, and other management practices.

Native Grazing Lands Are Widespread

The principal grazing lands are located in the Northern Plains, Mountain region, Pacific region, Southeast region, Delta region, and Southern Plains. Those in the western United States are generally grasslands or desert shrub lands too dry for arable farming. Likewise, mountain woodland, moist enough for tree growth but too steep for cultivation, accounts for a considerable acreage. Much of the mountain grazing land is federally owned. The principal native grazing lands of the South are composed of forested grazing lands, wet prairies, and cane marshlands. Nearly three-quarters of a billion acres of grazing land are involved in the first four named regions.

In addition, the grassland area in the Northern and Southern Plains regions comprises nearly a quarter of a billion acres. The livestock are grazed on a number of native short grasses, including grama grasses and buffalograss, with some tall grasses on the deep, sandy soils toward the eastern edge of the plains. The Flint Hills of eastern Kansas represent the only extensive area of native grazing land remaining in the humid tall-grass prairie region.

Of the approximately 100 million acres of open and forested grazing land in the Pacific Coast region, about 5 million acres are humid pastures of the northwestern part of this region and nearly 2 million acres are irrigated pastures.

Introduced Grasses and Legumes Are the Predominant Forage Species in the Lake States, Corn Belt, and Northeast Regions

Grazing in these three farm production regions is confined primarily to introduced plants from other parts of the world. These plants have been seeded into many pastures and have spread by their natural seed dispersion systems to adjoining land. In addition, the humid northwestern section of the Pacific region is also an important area where pastures have been improved by seeding introduced grasses and legumes which furnish a large share of the grazing.

Carrying Capacities of Pastures

The factors which are responsible for the wide variation in the carrying capacities of pastures are:

1. Variation in amount and distribution (seasonal) of moisture.
2. Variation in length of growing season (climatic).
3. Variation in growth characteristics of plants grazed.
4. Differences in fertility and moisture-supplying capacity of soils.

In many locations, 1 acre of good improved pasture will furnish all the herbage needed by a producing dairy cow or a beef cow and calf in a 12-months period. In other areas, 10 to 20 acres, or more, are required to meet the needs of one animal unit. See Fig. 1.3.

A study of these factors explains why so many of our highest producing pastures are located in the humid region of the South and along the coasts of Oregon and Washington.

In both of these locations, the growing season is long, and fertilizers and irrigation are used to supplement what nature provides. Likewise, irrigated pastures in the regions of long growing seasons of Arizona and California are very productive. It is a combination of these factors—i.e., long growing season, productive plant species, adequate moisture, and abundance of fertility—that makes pastures productive. All are essential. A deficiency of any one limits production.

Steep, rough land usually does not make the most productive pastures, regardless of fertility, simply because much of the water that falls on it runs off instead of percolating into the soil. This explains why the Blue Ridge areas of North Carolina and Georgia do not pro-

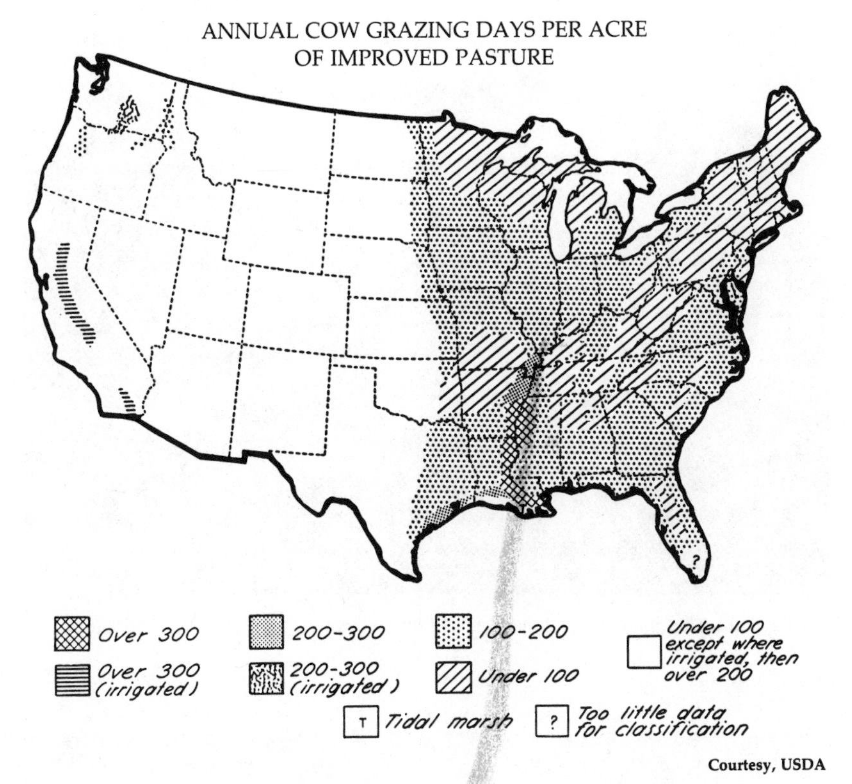

Courtesy, USDA

Fig. 1.3—Carrying capacity of pastures varies greatly.

vide a large amount of grazing per acre although the climate is well suited to production of pastures.

A New Concept

Today's progressive farmer knows that pasturage is one of the most valuable crops. This has not always been the case. For years, land not suited to cultivation, or land depleted of its natural fertility through continued cropping, was devoted to pastures. Far too often the steep, hilly, inaccessible acres were left to the livestock to climb over and forage what they could. In short, pastures were "step-children" of the farm.

Fields were "farmed" until crops were so poor they wouldn't pay

Fig. 1.4—The old concept: land too poor for cultivation is turned over to pastures.

Fig. 1.5—Beef cattle are among the principal users of pasture.

production costs, then were simply given over to pastures. Mostly, such fields were not even seeded to grass. Nature was supposed to provide herbage. Very few farmers thought they should add lime or fertilizer to acres devoted to pastures. Thus, as long as these theories prevailed, our "best" pastures were the hilly acres too steep or stony to plow.

As our agriculture advanced and agricultural research was more accepted, improvements in our nation's pastures were noticed. The old statement, "Nothing has happened to pastures in the past 25 years," no longer was true. Through the efforts of agricultural scientists and teachers, farmers gradually began to realize the blessings of improved pastures. Yet, unfortunately, even today there are many farmers who are not taking advantage of the profit and labor-saving advantages of productive pastures.

Good Pastures Are Profitable

Livestock programs that rely upon grazing for all or part of the herbage needs require a high-quality, highly productive pasture program to be most successful. Some dairy enterprises have adopted the total confinement system of livestock management. Their need

Fig. 1.6—Year-round grazing reduces livestock maintenance costs.

for pasture is nil or minor as for young stock or dry cows. Pastures still are an important feed source for dairy animals in the humid regions of the nation. Very productive grass-legume pastures are essential to the success of most livestock enterprises.

Productive pastures are vital for profitable sheep or beef production. Between 80 and 90 percent of the total annual feed needs of beef and sheep come from hay and pasture.

The hog farmer has improved production and reduced sanitation problems through the use of improved pastures.

Year-round grazing, popular in the South, has moved into the lower Corn Belt. Selecting forage species that make growth under cool temperatures, fertilizing, harvesting excess pasture, and stockpiling late summer pasture have made year-round grazing possible at more northern locations.

Pasture Programs for the Livestock Enterprise

Pasture improvement is a basic need on most livestock farms. Improvement can mean increased pasture production and increased quality. Such improvement provides for expansion of the livestock program and greater efficiency in the feeding of the livestock. More nutritious pastures can result in heavier weaning weights of calves born and reared on pasture, and higher conception rates of the cows. Re-breeding cows on a schedule that results in a calf every 12 months requires good herd management and good nutrition. Improved pastures can provide the high-quality feeds for the high nutritive plane needed by the cow.

High-quality pastures are valuable in the breeding and pregnancy period of other livestock such as sheep and horses. Pasture is greatly improved through proper fertilization, regular clipping, and the reseeding of improved species and varieties into low productive pastures.

The specific pasture program for a farm will depend upon several factors: class of livestock, geographic and climatic region, land values, yield potential of the soil, labor requirement, and management skills of the farm operator. Regions of high land values will require intensive pasture programs that produce high yields of high-quality herbage. Daily strip grazing a legume or a legume-grass pasture under irrigation is an example. Farmers in lower land value areas may choose

Fig. 1.7—Profitable beef enterprises need a calf from each cow every 12 months.

less intensive management-requiring species and fertility programs. An example is the rangelands of the West, where improved varieties of native range grasses are seeded onto existing rangelands.

Basic Pasture Improvement

Fertilization to increase production and encourage the increase of desired pasture plants *is* basic pasture improvement. Fertilization may include additions of nitrogen, phosphorus, potassium, and other essential elements. Lime is often needed in the humid regions of the United States to reduce soil acidity.

Many pastures of the humid regions of the United States have never been limed, and most are consequently acid. Desirable grasses and legumes do not thrive on very acid soils, and undesirable plants such as weeds, ferns, moss, and poverty grass have taken over. Liming such soils is the single most important step to improving these pastures. Additional phosphorus, potassium, and nitrogen will usually hasten the re-establishment of the desired pasture species.

Table 1.1

A Grazing Test Evaluation[1]

Factor	Check	Fertilized
Average inweight/head	367 lbs.	370 lbs.
Average stocking rates:		
Inweight/acre	147 lbs.	370 lbs.
Acre/head	2.5	1
Head/acre	0.4	1
Average gain/acre during test	60 lbs.	170 lbs.
Outweight/acre	207 lbs.	540 lbs.

[1]*Source:* California Agricultural Experiment Station.

Improved pastures will produce more herbage, and the herbage will usually be higher in nutritional value—higher in protein, digestible energy, and minerals—than unimproved pastures.

Pasture improvement by annual fertilization is well suited to lands that contain desirable pasture grasses but which are so steep or stony that tillage and reseeding by power machinery is hazardous or impossible.

Pasture Renovation

Pasture renovation is the establishment of one or more different species or varieties into an existing pasture. This may be accomplished through a tillage and reseeding procedure or by using herbicides and no-till seeding methods. With either method (including various modifications), high-yielding, high-quality legumes and/or grasses are seeded. Prior to tillage and/or seeding, the fertility needs are assessed by a proper soil test, and needed nutrients are applied.

No-till pasture renovation has become a well accepted practice from the Great Plains to the Humid Northeast. The procedures and equipment vary by regions. Consult your State Agricultural Experiment Station for latest techniques.

The Farmer Can Help

Infertile soils which have good moisture retention properties can be made into productive pastures by fertilization, reseeding, and grazing management. The soils of the southeastern United States are of this type. Generally, they are low in fertility and without treatment furnish only moderate grazing. On the other hand, when liberally fertilized and seeded to high-yielding grasses and legumes, these same soils become very productive pasture lands. Naturally the long growing season contributes to their high production.

Because of the favorable growing season and the ability of the soils to respond to treatment, the Southeast has become a significant livestock production center of our nation.

The mild climate in certain areas of the Pacific Northwest also contributes to a high carrying capacity of pastures in that area.

In the Northern Plains, Lake, and Northeastern regions, the shortness of the growing season limits the total production of the pastures. However, with the use of new, improved strains of grasses and legumes, coupled with the wise use of commercial fertilizer, manure, and lime, both the quantity and quality of the herbage can be greatly increased. In fact, with good cultural and management practices, farmers are learning that improved pasturage is their most profitable crop.

Courtesy, T.V.A.

Fig. 1.8—Fertility is essential for good growth. Airplanes are often used to apply fertilizer in the Great Plains Region and in mountainous regions.

It is these pastures that offer the greatest opportunities for increased income from good management, and therefore most of the discussion in this book will center around them. Generally these pastures are artificially made by preparing the seedbed and seeding to one or more of the highly productive grasses and legumes.

Subsequent chapters will deal with methods for improving the carrying capacities of pastures.

CHAPTER 2

Using Soils to Produce Pastures

Good land management consists of organizing and using all the land on the farm according to its capability. For lands in pasture this means proper land use with attention to erosion control, water conservation, maintenance of desirable soil reaction, addition of proper plant nutrients, organic matter replenishment, and, in some instances, rotation of crops.

Learn Characteristics of Soils

Proper land management requires some technical knowledge of soils and their characteristics. Following is a thumbnail sketch of desirable general information concerning soils.

1. Formation of soil.

Physical weathering (disintegration) and chemical weathering (decomposition) of the parent rock (rocks, minerals, and related materials at the earth's surface) provide size groups (separates) of gravel, sand, silt, and clay referred to as soil parent material.

By the action of climatic elements (principally temperature and rainfall) and biotic agents (microscopic and macroscopic plants and animals) on a particular parent material (of glacial, residual, or mixed origin) of a given topography (slope and drainage) for a period of time, a comparatively shallow layered profile, which is capable of supporting life, results.

Typically, a soil profile consists of surface and subsurface layers (A & A_1 horizons) which has decomposed organic material (humus)

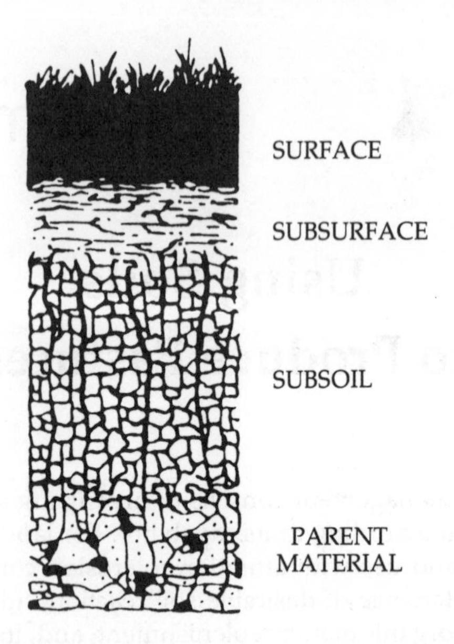

Fig. 2.1—A diagrammatic presentation of a soil profile.

mixed with the mineral matter, below which is a subsoil layer (B horizon) of considerable clay accumulation, and below which is a parent material layer (C horizon) of only slightly modified mineral material.

2. Components of soil.

On a volume basis, a silt loam surface horizon (the most widespread textural class) is comprised of 45 percent mineral matter, 5 percent organic matter (of which living organisms are a part), 25 percent water, and 25 percent air (the latter two being somewhat interchangeable).

3. Observable characteristics of soil.

Color—Usually the surface horizon (and subsurface in prairie soils) will be dark, moderately dark, light, or very light in color due to decreasing amounts of organic matter (humus). The subsoil (B horizon) may be dull, mottled, or bright due to increasing drainage and hence better oxidation of the iron compounds present in most soil bodies.

Texture—The size and proportion of the individual particles making up the mineral matter and organic matter (the solid body) of

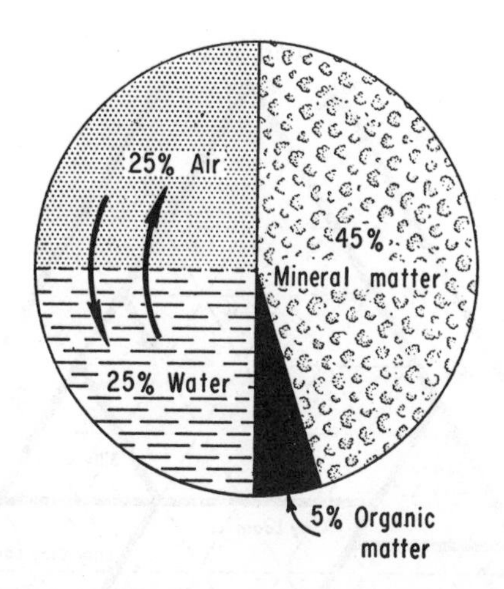

Fig. 2.2—Composition by volume of a silt loam soil in good condition for plant growth.

a soil are expressed in terms of sand, silt, and clay percentages. See Fig. 2.3 for a textural class chart.

Fine textured—clay, silty clay, sandy clay.

Moderately fine textured—silty clay loam, clay loam, sandy clay loam.

Coarse textured—loamy sand, sandy loam.

Very coarse textured—gravelly loam, sand.

Organic—peat (almost completely organic material), muck (some mineral material incorporated with well-decomposed organic material).

In the field, the texture (soil textural class) is determined by squeezing a moist sample of the soil between the thumb and forefinger, while in the laboratory, a technician determines textural class by running a mechanical analysis.

Structure—Due to the binding effect of humus and perhaps the "bonding" effect of multi-valent elements in the mineral matter portion of soils, aggregates of individual particles occur to result in granular, platy, blocky, or single grain (sand) structure of surface soils and prismatic or massive structure of subsoils. See Fig. 2.4.

Permeability—The size of the pores and the amount of porosity determine the readiness or slowness of movement of air or water in a soil body.

Moisture-holding capacity—The amount of humus and the tex-

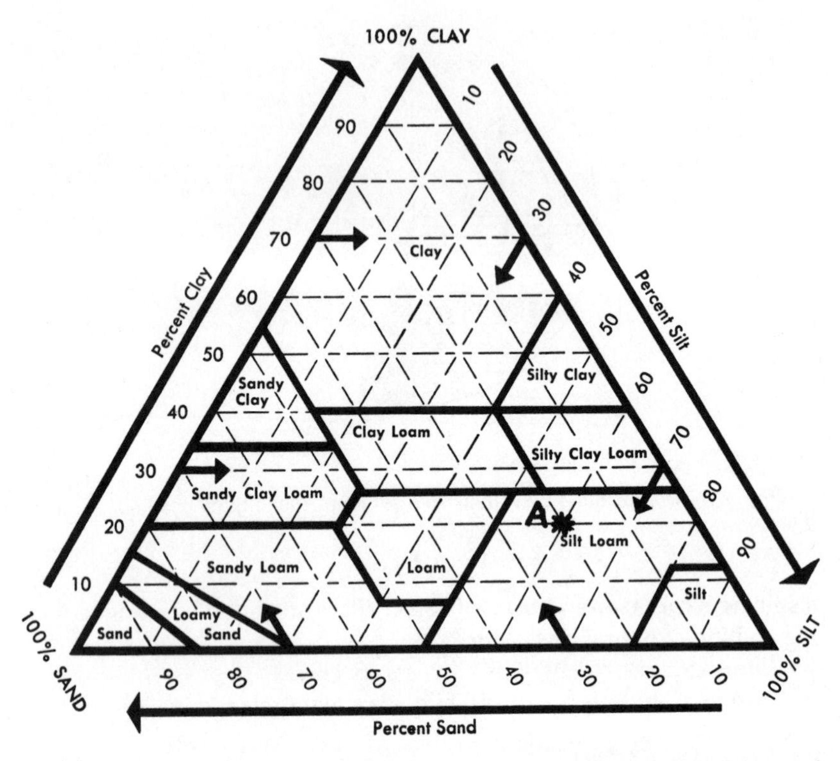

Fig. 2.3—This textural class chart gives the proportions of the separates in some of the classes commonly used in describing soils. The sum of the percentages of sand, silt, and clay at any point in the triangle is 100. Point A represents 20 percent sand, 20 percent clay, and 60 percent silt. The textural class name of this soil is silt loam.

ture, structure, and depth of soil determine its water-holding capacity.

Depth of soil—The structure and inherent fertility largely determine depth of soil favorable for root growth, deep (36 inches or more) to shallow (less than 10 inches).

Shape—The areal extent (boundaries) of a particular surface textural soil body, including slope, is referred to as shape.

Slope—This refers to the degree to which the observable terrain of the surface of the textural soil body is depressional, level, undulating, gently sloping, rolling, steep, or very steep.

Degree of erosion—This ranges from deposition (+) through no apparent erosion (0), slight (1), moderate (2), severe (3), very severe (4), and very severe gullies (5). The last category includes areas where

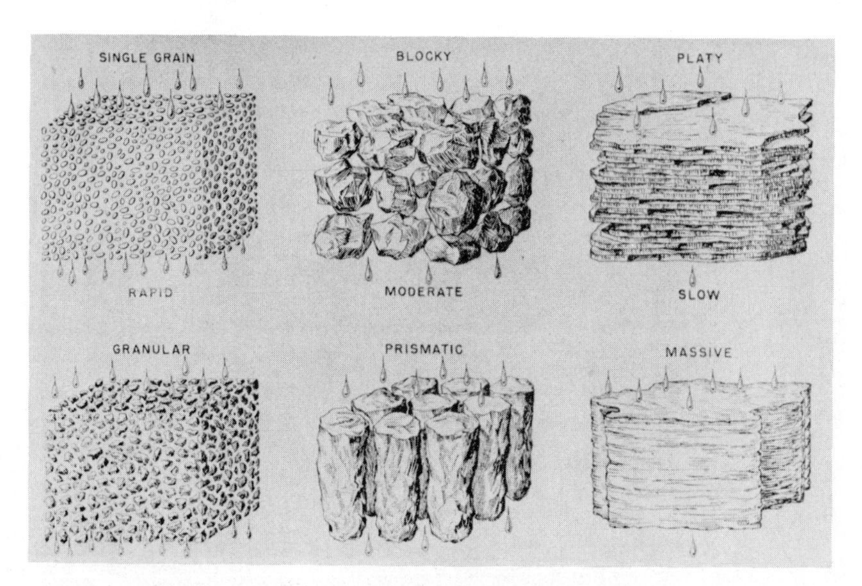

Fig. 2.4—Kinds of soil structure and moisture movement.

deep gullies or blowouts have ruined the land for agricultural purposes.

4. Capability and suitability of land for different uses (land use capability classes).

The term "land use capability" is used to designate a classification of land which is indicative of its potential productivity. The type of soil, percent of slope, and amount of surface horizon present or the degree of erosion are used in determining the capability class of the land. Such a classification does not take into consideration soil fertility. It is indicative of potential productivity only if a program of soil conditioning and fertility maintenance accompanies the use of the land in accordance with the capability classification.

Soil conservationists have divided land into eight broad classes on the bases of capability and suitability for different uses. Land in the first four classes can be cultivated, while land in the last four classes needs a permanent cover of vegetation.

The trend has been to use Class III land or higher for pastures, but with the move to get some cropland out of intertilled crops, some Class II and perhaps Class I land may shift over to pasture production, as an extension of conservation measures.

Courtesy, USDA

Fig. 2.5—In this landscape, each of the eight capability classes of land is identified.

The following paragraphs indicate general conditions and a brief summary of possible uses for each class.

Class I land

Class I land is suitable for cultivation without the application of special conservation practices. It must be workable, free of stones, and productive enough, considering climate, to give at least moderate yields of the common crops. It must be nearly level, not subject to more than slight erosion, free from overflows that interfere with crop production, and sufficiently well drained to permit adequate production. Land in this class may, and probably will, require the application of fertilizers and limestone and the rotation of crops, including a green manure crop to maintain soil structure and to insure continued average yields. Such land is usually the most desirable cropland and is frequently, though not always, the most productive. It is well suited to growing row crops.

Class II land

Class II land is suitable for cultivation with the application of simple conservation practices. Such practices may be required to control erosion, conserve water, or drain wet soils. Contour farming, strip cropping, cover crops, and crop rotations which include grasses and legumes may be methods required to conserve soil and water on Class II land.

Class III land

Class III land is suitable for permanent cultivation only if intensive conservation methods are used. Such land usually requires that a combination of the best possible conservation practices be maintained on the land to insure continued production. In addition to the methods used on Class II land, Class III land may require terraces, grassed waterways, and the extension of crop rotations to include more years in meadow and small grain with fewer years in row crops.

Class IV land

Class IV land is suited to only occasional or limited cultivation. It may be steeper than Class III land, more severely eroded, more susceptible to erosion, or more difficult to drain. It is not well suited to cultivation and is best used for permanent vegetation.

Class V land

Class V land is not suitable for cultivation but must remain under permanent vegetative cover. Land in this class is largely soil that is too wet or too stony for cultivation. It may be used for grazing purposes or for woodland with good management practices.

Class VI land

Class VI land is suitable for grazing or woodland under permanent vegetative cover with only moderate restrictions on use. It is subject to severe erosion and should never be cultivated. Usually such land requires reseeding and the application of limestone and fertilizers before it can be used for grazing.

Class VII land

Class VII land is not suitable for cultivation and requires severe restrictions if it is used for pasture or woodland. In addition to reseeding and the application of lime and fertilizers, careful regulation of graz-

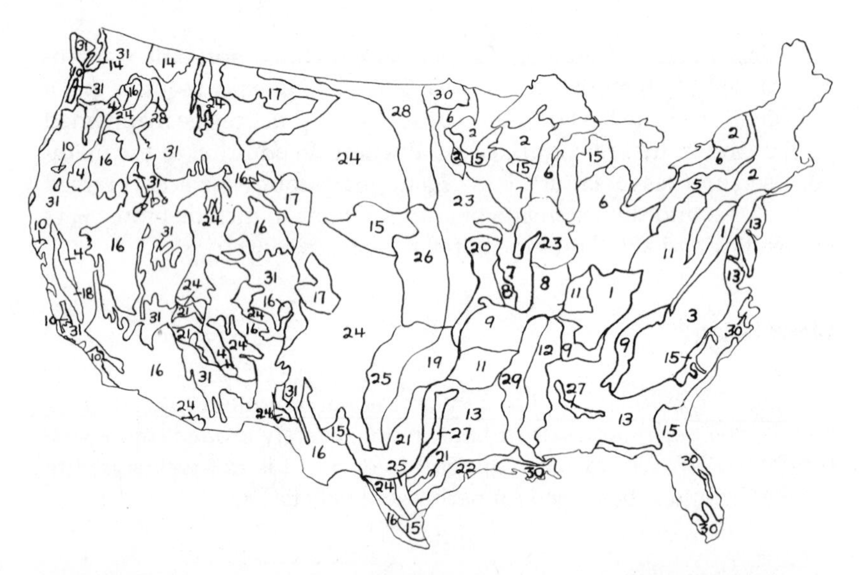

1. Brown silt loams
2. Brown gravelly and stony loams
3. Reddish sand and clay loams
4. Red-brown soils
5. Grayish to brownish silt loams, often poorly drained
6. Light brown silty loams
7. Gray to brown silty soils, compact silt-clay subsoils
8. Gray to brown soils, heavy clay subsoils
9. Yellowish to reddish silt loams
10. Yellowish brown soils
11. Yellowish silty sand to stony loams
12. Yellowish silt loams
13. Yellowish sandy loams
14. Light-colored soils
15. Sand resting on clay
16. Gray or brown arid soils
17. Medium dark soils, calcareous subsoils
18. Brown soils of Pacific valleys
19. Dark brown soils
20. Dark brown loams, heavy brown subsoils
21. Shallow stony soils from limestone
22. Dark, calcareous soils, poorly drained
23. Dark brown silty soils
24. Chestnut brown soils, calcareous subsoils
25. Dark brown soils, calcareous subsoils
26. Very dark brown soils, calcareous subsoils
27. Black soils, highly calcareous
28. Black soils, highly calcareous subsoils
29. Alluvial soils
30. Marsh and swamp
31. Rough and mountainous

Courtesy, Western Illinois University

Fig. 2.6—Map of soil textural regions of the United States.

ing is necessary. Most Class VII land is recommended for use as woodland rather than pasture. Because this land is subject to erosion, careful management of timber is necessary.

Class VIII land

Class VIII land is suited neither to cultivation nor to the production of useful permanent vegetation. It is chiefly rough, stony, barren land or marshes that are permanently wet and cannot be drained. Some Class VIII land, particularly the swamps and marshes, may be made to produce a profitable crop of wildlife. Other areas may be most useful as recreational and scenic areas.

The analysis of the land included in a farm and the development of a land use capability map of the farm are the basis upon which the farm conservation plan is made. The farm conservation plan is designed to make the best possible use of the land under the type of farming the land operator wishes to carry on.

After the land has been classified, the Soil Conservation Service

Fig. 2.7—Well drained, adequately fertilized soils supply excellent forage of various types.

technician and the farmer work out the details of operation, determining what changes are necessary in field arrangement, crop rotation, fertility maintenance, livestock, etc. In some instances this may necessitate a rather radical change in operation, a change which will be made gradually over a period of years. In others it may require little variation from present practices and can be accomplished easily and quickly. In all instances it is the farmer, taking advantage of consultation with the technician, who decides what changes are to be made and when and how these changes are to be accomplished. The result will be a farming operation suited to the farmer which will provide the highest possible continuing income from the land.

CHAPTER 3

Pasture as an Economical Source of Livestock Feed

Most agricultural experts agree that pastures provide Nature's most perfect livestock feed at the lowest possible cost. Young, succulent pasture is high in protein and rich in minerals and vitamins, and it is relished by all classes of livestock.

Activities Which Involve Approved Practices

1. Selecting pasture plant species suited to soil and climatic conditions.
2. Improving pastures to produce large amounts of feed per acre.
3. Saving labor by use of livestock.
4. Managing pastures to improve soil productivity.
5. Growing pastures on steep and rolling land to prevent loss of soil by erosion.

1. Selecting Pasture Plant Species Suited to Soil and Climatic Conditions

There are high-producing grasses and legumes adapted for nearly every section of the country. The secret of making pastures pay off lies in using the right mixtures, keeping the herbage growing luxuriantly, and harvesting it at the proper stage.

Fig. 3.1—A profitable combination of good pasture and high-producing livestock.

Plant a combination of grasses and legumes

This means that the farmer should use a combination of grasses and legumes that will provide the livestock with good grazing throughout the growing season. It also reduces the chances of bloat while grazing on pastures.

2. Improving Pastures to Produce Large Amounts of Feed per Acre

Nearly every state agricultural experiment station has developed data which show how efficient pastures are. Typical examples are given in this chapter.

Use pastures for cheapest food

When one compares the production of total digestible nutrients (TDN) and protein from grasses with that of corn, it becomes apparent that, if properly fertilized and harvested, forages or pastures can be valuable. Researchers working in northern Minnesota made this comparison. These trials demonstrated that fertilized mixed grasses produced 3,500 pounds of total digestible nutrients per acre. This was equivalent to a 78-bushel-per-acre corn yield, and the possibility of producing 78-bushel corn in northern Minnesota on the land used for pasture was rather unlikely. The grasses produced more than 2¼ times the protein of the 78-bushel corn crop. Complete data are presented in Table 3.1.

From other research in Minnesota, scientists have calculated the comparative cost of producing TDN from various feed sources. These data are presented in Table 3.2.

Cornell University gives these figures for New York. The relative cost of 100 pounds of total digestible nutrients from pasture, 100; hay, 200; corn silage, 333; grains and commercial by-products feeds, 666.

Fig. 3.2—Even a corn crop like this cannot compete with good improved pasture as an economical feed producer.

Table 3.1

Production of TDN and Protein from Grass Compared to Corn[1]

	Tons Hay/Acre	Lbs. TDN/Acre	Lbs. Protein/Acre
Permanent bluegrass (no fertilizer)	0.8	800	160
Permanent bluegrass (+ fertilizer)	2.0	2,000	600
Mixed grasses (± fertilizer)	3.5	3,500	1,000
Corn (bu.)	18[2]	800	90[3]
Corn (bu.)	45[2]	2,000	227[3]
Corn (bu.)	78[2]	3,500	393[3]

[1]*Source:* Charles Simkins, University of Minnesota.
[2]Equivalent.
[3]Based on 9% protein in corn.

Table 3.2

Relative Feed Cost per 100 Pounds of TDN

Fertilized grass pasture	100
Alfalfa brome pasture	90
Alfalfa brome silage	217
Alfalfa brome hay	217
Oat silage	251
Corn silage	300
Corn grain	349
Oats grain	396
16% dairy concentrate	452

Extend the grazing season

Information on pasture and forage varieties for special conditions within an area is available from the state Agricultural Experiment Station, county agricultural agents, teachers of vocational agriculture, and specialists of the Soil Conservation Service. The reader is advised to check with the foregoing for variety recommendations for the proposed usage in his or her area or region.

Table 3.3 gives a listing of annual and perennial grasses and legume plants for each farm production region from among which a pasture calendar of grazing may be set up.

Table 3.3

Calendar of Pasture Grasses and Legumes for the Production Regions

Pasture Crop	Spring	Late Spring & Early Summer	Late Summer & Early Fall	Late Fall & Early Winter
APPALACHIAN REGION				
Bromegrass		xxxxxxxxxxxxxxxxxxxxxxx	xxxxxxxxxxxxxxxxxxxxxxxxx	
Tall fescue	xxxxxxxxxxxxxxxxxxxxxxxx		xxxx xxxxxxxxxxxxxxxxxxxxxxxx	
Orchardgrass	xxxxxxxxxxxxxxxxxxxxxxxxxxxxx		xxxxx xxxxxxxxxxxxxxxxxxxxxxxxxxxx	
Timothy		xxxxxxxxxxxx		xxxxxxxxx
Small grains (fall seeded)	xxxxxxxxxxxxxxxxxxxxxxxxxxxxxxx			
Pearl millet or sudangrass			xxxxxxxxxxxxxxxxxxxxxxxxx	
Small grains				xxxxxxxxx
Bermudagrass		xxxxxxxxxxxxxxxxxxxxxxxxxxxxxxxxxxxx		
Ryegrass		xxxxxxxxxxxxxxxxxxxxxxx		
Kentucky bluegrass	xxxxxxxxxxxxxxxxxxxxxxxx		xxxxxxxxxxxxxxxxxxxxxxx	
Reed canarygrass		xxxxxxxxxxxxxxxxxxxxxx	xxxxx xxxxxxxxxxxxxxxxxxxxxx	
Switchcane (North Carolina Coastal Plain)	xxxxxxxxxxxxxxxxxxxxxxxx		xxxxxxxxxxxxxxxxxxxxxxxxxxxxxxx	
Red clover		xxxxxxxxxxxxxxxxx	xxxxx xxxxxxxxxx	
Alsike clover		xxxxxxxxxxxxxxx	xxxxxxxxxxxxx	
Sweet clover		xxxxxxxxxxxxxxxxx	xxxxxxxxxx	
Alfalfa		xxxxxxxxxxxxxxxxxxxxxxx	xxxxxxx xxxxxxxxxxxxxxxxxxxxxxx	
Ladino clover		xxxxxxxxxxxxxxx	xxxxxxxxxxxxxxxxxx	
White clover		xxxxxxxxxxxxxxxxx	xxxxxxxxxxxxxxxxxxxxxxxxxxx	
Crimson clover		xxxxxxxxxxxxxxxxxxxxx	xxxxxxxxxxxxxxxxxxx	
Perennial lespedeza	xxxxxxxxxxxxxxxxxxxxxxxx			
Annual lespedeza		xxxxxxxxxxxxxxxxxxxxxxxxxxxxxxxxxxxxxxx		
Hairy vetch	xxxxx			xxxxxxxx

(Continued)

Table 3.3 (Continued)

Pasture Crop	Spring	Late Spring & Early Summer	Late Summer & Early Fall	Late Fall & Early Winter
SOUTHEASTERN REGION				
Ryegrass	x	x		
Coastal bermudagrass		x	x	x
Tall fescue	x	x		x
Pearl millet or sudangrass			x	x
Johnsongrass (may become a pest)		x	x	
Bahiagrass		x	x	
Dallisgrass		x	x	
Pangola (Peninsular Florida)		x	x	x
Small grains (winter or fall seeded)	x	x		x
Perennial lespedeza	x	x		
Annual lespedeza		x	x	
Ladino clover		x		x
White clover		x	x	x
Crimson clover	x	x		
Red clover		x	x	
Alfalfa (where adapted)		x	x	
SOUTHERN PLAINS (Uplands with adequate natural or supplied moisture)				
Ryegrass		x		x
Bromegrass		x		x
Meadow fescue	x	x		x
Small grains	x	x		x
Sudangrass		x	x	
Blue panicgrass		x	x	
Sweet clover		x	x	x
Annual lespedeza		x	x	
Red clover		x	x	x
White clover (where adapted)	x	x	x	x

NORTHERN PLAINS

Bromegrass	xxxxxxxxxxxxxxxxxxxxxxxxxx		xxxxxxxxxxxxxxxxxxxxx
Crested wheatgrass	xxxxxxxxxxxxxxxxxxxxxxxx		xxxxxxxxxxxxxxxxx
Intermediate wheatgrass	xxxxxxxxxxxxxxxxxxxxxx		xxxxxxxxxxxxxxxxxxxxx
Russian wildrye	xxxxxxxxxxxxxxxxxxxxx		xxxxxxxxxxxxxxxxxx
Timothy	xxxxxxxxxxxxx		xxxxxxxxxxxxxxxxxxxxx
Sudangrass		xxxxxxxxxxxxxxxxxxxxxxxxxxxxxxxx	
Small grains	xxxxxxxxxxxxxxxxxxxxxxxxxx		
Orchardgrass	xxxxxxxxxxxxxxxxxxxxxxxxxxxxxx	xxxx	xxxxxxxxxxxxxxxxxxxx
Reed canarygrass	xxxxxxxxxxxxxxxxxx	xxxx	xxxxxxxxxxxxxxxxxxxxxxx
Native grasses (Slender wheatgrass, Buffalo grass, bluestem, switchgrass)	xxxxxxxxxxxxxxxxxxxxxxxxxxxxxxxxxxxxxxxxxxxxxx		
Alfalfa	xxxxxxxxxxxxxxxxxxxxxxxxxxxxxxxxxxxxxxxxxxxxxxxxxxxx		
Sweet clover	xxxxxxxxxxxxxxxxxxxxxxxxxxxxxxxxxxxxxx		
Red clover	xxxxxxxxxxxx		xxxxxxxxxxxxxxxx

CORN BELT REGION

Bromegrass	xxxxxxxxxxxxxxxxxxxxxxxxxxxx		xxxxxxxxxxxxxxxxxxxxxxxxxxxxxxxxxxxx
Tall fescue	xxxxxxxxxxxxxxxxxxxxxxxxxxxxxxxxx	xxxx	xxxxxxxxxxxxxxxxxxxxxxxxxx
Orchardgrass	xxxxxxxxxxxxxxxxxxxxxxxxxxxxxx	xxxxx	xxxxxxxxxxxxxxxxxxxxxxxxxxx
Small grains (rye)	xxxxxxxxxxxxxxx		xxxxxxxxxxxxxxxxxxx
Timothy	xxxxxxxxxxxxxxxxx		xxxxxxxxxxxxxxxxxxxxxxx
Kentucky bluegrass	xxxxxxxxxxxxxxxxxxxxxxxxxxxxxxx		xxxxxxxxxxxxxxxxxxxx
Redtop	xxxxxxxxxxxxxxxxxx		xxxxxxxxxxxxxxxxxxxxx
Reed canarygrass	xxxxxxxxxxxxxxxxxxxxxxxx		xxxxxxxxxxxxxxxxxxxxxxxxxxxx
Alfalfa	xxxxxxxxxxxxxxxxxxxxxxxxxxxxxx	xxxxx	xxxxxxxxxxxxxxxxxxx
Red clover	xxxxxxxxxxxxxxxxxxxxxxx	xxx	xxxxxxxxxxxxxxxxxxxxx
Sweet clover	xxxxxxxxxxxxxxxxxxxxxxxxx		xxxxxxxxxxxxxxxxxxx
Ladino clover	xxxxxxxxxxxxxxxxxxxxxx		xxxxxxxxxxxxxxxxx
Birdsfoot trefoil	xxxxxxxxxxxxxxxxxxxxxxx	xxxxx	xxxxxxxxxxxxxxxxxxx
Sudangrass		xxxxxxxxxxxxxxxxxxxxxxxxx	

(Continued)

Table 3.3 (Continued)

Pasture Crop	Spring	Late Spring & Early Summer	Late Summer & Early Fall	Late Fall & Early Winter
NORTHEASTERN REGION				
Orchardgrass		xxxxxxxxxxxxxxxxxxxxxxxxxxxx	xxxx xxxxxxxxxx	
Timothy		xxxxxxxxxxxxxxxxx	xxxxxxxxxxxx	
Reed canarygrass		xxxxxxxxxxxxxxxxxxxxxxxxxxxxxxxx	xxxx xxxxxxxxxxxxxxxxxxxx	
Bromegrass		xxxxxxxxxxxxxxxxxxxxx	xxxxxxxxxxxxxxxxxxx	
Small grains (fall seeded)		xxxxxxxxxxxxxxxxxxxxxxxxxx	xxxxxxxxxxxxxxxxx	
Kentucky bluegrass		xxxxxxxxxxxxxxxxxxx	xxxxxxxxxxxxxxxxxxxxxx	
Alfalfa		xxxxxxxxxxxxxxxxxxxxxxxxx	xxx xxxxxxxxxxxxxxxx	
Birdsfoot trefoil		xxxxxxxxxxxxxxxxxxxxxxx	xxx xxxxxxxxxxxxxxxxxxx	
Ladino clover		xxxxxxxxxxxxxxxxxxxxxx	xxxx xxxxxxxxxxxxxxxx	
White clover		xxxxxxxxxxxxxxxxxxxxxxxxxxxxx	xxxxxxxxxxxxxx	
Red clover		xxxxxxxxxxxxxxxxxxxx	xxx xxxxxxxxxx	
Alsike clover		xxxxxxxxxxxxxxxxxxxxx	xxxxxxxxxxxx	
Annual lespedeza		xxxxxxxxxxxxxxxxxxxxxxxxxxxxxxxxxxxxxxxxxxxxxxxx		
DELTA STATES REGION				
Coastal bermudagrass		xxxxxxxxxxxxxxxxxxxxxxxxxxxxxxxxxxxxxxxxxxxxxx		
Orchardgrass		xxxxxxxxxxxxxxxxxxxxxxxxxx	xxxxxxxxxxxxxxxxxxxxxxxxxxxxx	
Tall fescue		xxxxxxxxxxxxxxxxxxxxxxxxxx	xxxxxxxxxxxxxxxxxxxxxxxxxxxx	
Small grains	xxxxxxxxxx			xxxxxxxxxxxxx
Sudangrass or millet		xxxxxxxxxxxxxxxxxxxxxxxxxxxxxxxxxxxxx		
Dallisgrass		xxxxxxxxxxxxxxxxxxxxxxxxxxxxxxxxxxxxxxxxxxxxxxxxx		
Bahiagrass		xxxxxxxxxxxxxxxxxxxxxxxxxxxxxxxxxxxx		
Ladino clover		xxxxxxxxxxxxxxxxxxxxxx	xxxxxxxxxxxxxxxxxxxxx	
Crimson clover	xxxxxxxxxxxxxxxxxxxxxxxxxxxx			
Perennial lespedeza	xxxxxxxxxxxxxxxxxxxxxxxxxxxx			
Annual lespedeza		xxxxxxxxxxxxxxxxxxxxxxxxxxxxxxxxxxx		

LAKE STATES REGION

Bromegrass	xxxxxxxxxxxxxxxxxxx	xxxx	xxxxxxxxxxxxxxxxxxxxxxxxxxxxx
Orchardgrass	xxxxxxxxxxxxxxxxxxxxx	xxxxx	xxxxxxxxxxxxxxxxxxxxxxxxx
Kentucky bluegrass	xxxxxxxxxxxxxxxxxxx		xxxxxxxxxxxxxxxxxxxxxxxxxx
Timothy	xxxxxxxxxxxx	xx	xxxxxxxxxxxxxxxxxxxxxxxxx
Reed canarygrass	xxxxxxxxxxxxxxxxx	xxxxx	xxxxxxxxxxxxxxxxxxxxxxxxxxxxxx
Small grains	xxxxxxxxxxxxxxx		xxxxxxxxxxxxxxxxxxxxxx
Sudangrass		xxxxxxxxxxxxxxxxxxxxxxxx	
Meadow fescue (if adapted)	xxxxxxxxxxxxxxxxxxxxxxxxxxxxxxxxxx		xxxxxxxxxxxxxxxxxxxxxxxxxxxxx
Alfalfa	xxxxxxxxxxxxxxxxxxxxxxxxx	xxxx	xxxxxxxxxxxxxx
Ladino clover	xxxxxxxxxxxxxxxxxxxxxxx	xxx	xxxxxxxxxxxxxxxxxxxx
Red clover	xxxxxxxxxxxxxxxxxxxxxxx	xxx	xxxxxxxxxxxxxxxxx
Birdsfoot trefoil	xxxxxxxxxxxxxxxxxxxxxxxxxxxxxx	xxx	xxxxxxxxxxxxxxxx
Annual lespedeza			xxxxxxxxxxxxxxxxxxxxxxxxxxx
Alsike clover	xxxxxxxxxxxxxxxxxxxxxxxxxxxxxxxxx		xxxxxxxxxxxxxxxxxxxxxxx
Sweet clover	xxxxxxxxxxxxxxxxxxxxxx		xxxxxxxxxxxxxxxxxxxxxx

PACIFIC REGION
(Where moisture is adequate or supplied)

Ryegrass	xxxxxxxxxxxxxxxxxxxxxxxxxxxxxxxxx		xxxxxxxxxx
Smooth bromegrass	xxxxxxxxxxxxxxxxxx	xxx	xxxxxxxxxxxxxxxxxx
Orchardgrass	xxxxxxxxxxxxxxxxxxxx	xxxx	xxxxxxxxxx
Tall fescue	xxxxxxxxxxxxxxxxxxxxx		xxxxxxxxxx
Wheatgrass	xxxxxxxxxxxxxxxxxxxx		xxxxxxxxx
Bluegrass	xxxxxxxxxxxxxxxxxx		xxxxxxxxx
Reed canarygrass	xxxxxxxxxxxxxx	xxx	xxxxxxxxxxxxxxxxx
Timothy (where adapted)	xxxxxxxxxxxx		xxxxxxxxxx
Small grains	xxxxxxxxxxxx		xxxxxxxxxx
Tall native grasses	xxxxxxxxxxxxxx	xxxxxx	
Short native grasses			xxxxxxxxxx
Alfalfa	xxxxxxxxxxxxxxxxxxxx		xxxxxxxxxxxxxxxxx
Sweet clover	xxxxxxxxxxxxxx		xxxxxxxxxxxxx
Ladino clover	xxxxxxxxxxxxxxxxxxxx		xxxxxxxxxxxxxxxxxxx
Alsike clover	xxxxxxxxxxxxxxxxxxxx		xxxxxxxxxxxxxxxxx
Big trefoil (coastal)	xxxxxxxxxxxxxxxxxx		xxxxxxxxxxxxx
Sudangrass			xxxxxxxxxxxxxxxxxxxxxxxx

(Continued)

Table 3.3 (Continued)

Pasture Crop	Spring	Late Spring & Early Summer	Late Summer & Early Fall	Late Fall & Early Winter
MOUNTAIN (INTERMOUNTAIN) REGION				
Meadow foxtail	xxxx	xxxxxxxxxxxxxxxx	xxxxxxxxxxxxxxxxxxxxxxxxxxxxxxxx	
Timothy		xxxxxxxxxxxxxxxx	xxxxxxxxxxxxxxx	
Smooth bromegrass	xxxx	xxxxxxxxxxxxxxxx	xxxxxxxxxxxxxxxxx	
Intermediate wheatgrass		xxxxxxxxxxxxxxxxxxxxxx	xxxxxxxxxxxxxxx	xxxx
Russian wildrye	xxxxxxxx	xxxxxxxxxxxxxxxx	xxxxxxxxxxxxxxxxxxxxxx	
Reed canarygrass		xxxxxxxxxxxxxxxxxxxxxxxxxx	xxxxxxxxxxxxxxx	
Orchardgrass (if will withstand winters)		xxxxxxxxxx	xxxxxxxxxxxxxxxxxxxxxx	
Alfalfa		xxxxxxxxxxxxxxxxxxxxxxx	xxxxxxxxxxxxxxxxxxxx	
Alsike clover		xxxxxxxxxxxxxxxxxxxx	xxxxxxxxxxxxxx	xx
Birdsfoot trefoil (if will withstand winters)		xxxxxxxxxxxxxxxx	xxxxxxxxxxxxxxxx	xx
Native short and mid grasses	xxxx	xxxxxxxxxx	xxxxxxxxxxxx	xx
RANGELAND (DRY) PASTURE CALENDAR (Northern Shortgrass Plains)				
Crested wheatgrass	xxxxxxxx	xxxxxxxxxxxxxxxx		
Blue grama			xxxxxxxxxxxxxxxxxxxxx	xxxxxxxxxxxxxxxxxx
Little bluestem		xxxxxxxxxxxxxxxx		
Buffalograss			xxxxxxxxxxxxxxxxxxxx	xxxxxxxxxxxx
Sideoats grama			xxxxxxxxxxxxxxxxx	xxxxxxxxxxxxxxxxxx
Needle and thread grass	xxxxxxxx			
Kentucky bluegrass		xxxxxxxxxxxxxxxx	xxxxxxxxx	xxxxxxxxxx
Western wheatgrass	xxxxxx	xxxxxxxxxxx	xxxxxxx	xxxxxxxxxxxx

RANGELAND (DRY) PASTURE CALENDAR

(Southern Shortgrass Plains)

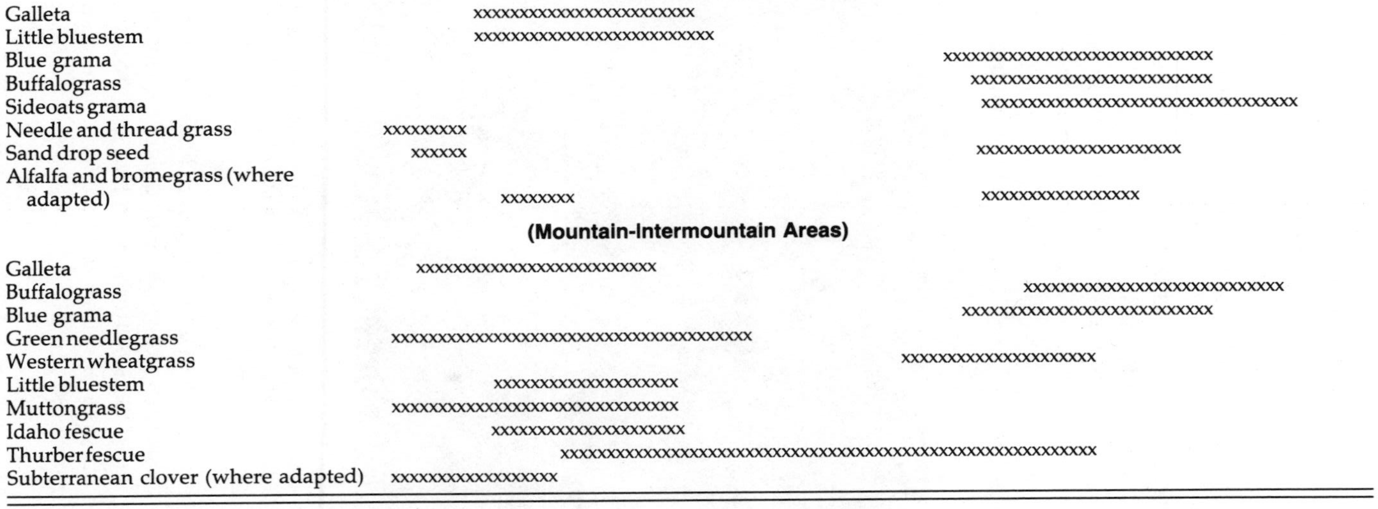

Galleta	xxxxxxxxxxxxxxxxxxxxxxxx	
Little bluestem	xxxxxxxxxxxxxxxxxxxxxxxxxx	
Blue grama		xxxxxxxxxxxxxxxxxxxxxxxxxxxxx
Buffalograss		xxxxxxxxxxxxxxxxxxxxxxxxxx
Sideoats grama		xxxxxxxxxxxxxxxxxxxxxxxxxxxxxxxxx
Needle and thread grass	xxxxxxxxx	
Sand drop seed	xxxxxx	xxxxxxxxxxxxxxxxxxxxxx
Alfalfa and bromegrass (where adapted)	xxxxxxxx	xxxxxxxxxxxxxxxxx

(Mountain-Intermountain Areas)

Galleta	xxxxxxxxxxxxxxxxxxxxxxxxxx	
Buffalograss		xxxxxxxxxxxxxxxxxxxxxxxxxxxx
Blue grama		xxxxxxxxxxxxxxxxxxxxxxxxxxx
Green needlegrass	xxxxxxxxxxxxxxxxxxxxxxxxxxxxxxxxxxxxxxx	
Western wheatgrass		xxxxxxxxxxxxxxxxxxxxx
Little bluestem	xxxxxxxxxxxxxxxxxxxx	
Muttongrass	xxxxxxxxxxxxxxxxxxxxxxxxxxxxxxx	
Idaho fescue	xxxxxxxxxxxxxxxxxxxxx	
Thurber fescue	xxxxxxxxxxxxxxxxxxxxxxxxxxxxxxxxxxxxxxxxxxxxxxxxxxxxxxxxxx	
Subterranean clover (where adapted)	xxxxxxxxxxxxxxxxxx	

3. Saving Labor by Use of Livestock

One of the real advantages of good pasture is that it reduces the labor requirements on the farm. This is true because the livestock harvest the crop and no direct labor is required by the farmer to process the crop and feed it to the animals.

The Ohio Agricultural Experiment Station has found that it takes 12 percent less labor to do the daily chores when the dairy herd is on pasture than when it is barn-fed. The time required for taking the cows to and from the pasture was included in the study.

Courtesy, USDA

Fig. 3.3—Animals do the harvesting. With proper management, pastures can be put to multiple use to include grazing, wildlife habitat, and recreation areas.

The North Carolina Agricultural Experiment Station has calculated that the return per work-hour of labor spent in establishing and managing improved pastures brought a 4:1 greater return than from wheat, a 6.5:1 greater return than from corn, and an 8.6:1 greater return than from oats.

4. Managing Pastures to Improve Soil Productivity

The words "improved pastures" are almost synonymous with perfect soil and water conservation. Once the pasture is established, the dense foliage intercepts the pounding action of raindrops, destroying their soil-deteriorating power. As a result, the water trickles into the

Courtesy, USDA

Fig. 3.4—Dense foliage intercepts and destroys the harmful effect of raindrops.

soil rather than the soil aggregates being broken down to individual soil grains and then being washed away. In addition, the dense root growth also helps hold the soil in place and offers further resistance to the lateral movement of water.

5. Growing Pastures on Steep and Rolling Land to Prevent Loss of Soil by Erosion

Use manure to prevent soil depletion

Livestock droppings return to the soil a considerable portion of the plant food nutrients contained in the herbage. In an experiment at the New Jersey Dairy Research Farm at Sussex, it was found that a 1,300-pound Holstein cow produced annually about 21 tons of manure. Of this, 25 percent was urine and 75 percent feces. Each ton of the mixed excrement carried 9.5 pounds of nitrogen, 3 pounds of phosphate, and 8 pounds of potash. Calculations showed that 70 percent of the nitrogen in the feed, 63 percent of the phosphate, and 86 percent of the potash in the feed passed through the animal and was returned in the manure.

Cornell University has published information on the average composition of different kinds of livestock manure. This information is given in Table 3.4. The Cornell figures for cow manure are remarkably similar to the New Jersey data.

Fortunately for humanity, pastures provide a high-yielding crop that can be grown on rolling land which suffers severe soil erosion when under cultivation. Many of these areas abound in good stands

Table 3.4

Plant-Food Nutrients per Ton of Manure, Including Solid, Liquid, and Bedding[1]

Kind of Animal	Pounds Nitrogen (N)	Pounds Phosphate (P_2O_5)	Pounds Potash (K_2O)	Tons Manure Produced per Year per 1,000 Lbs. Live Wt.
Cow	11.4	3.1	9.9	15
Duck	11.4	28.8	19.8	4½
Hen	21.0	16.4	10.2	4½
Horse	13.2	5.1	12.1	12
Pig	9.9	6.7	9.3	18¼
Sheep	15.8	6.7	18.0	9¾
Steer	15.0	6.0	8.0	9

[1]*Source:* Cornell University.

of native grasses. Others can be successfully seeded to mixtures of high-producing grasses and legumes. Approved methods for establishing such mixtures are given in other chapters.

Reduce plowing

Modern science is now demonstrating how soil losses which sometimes occur on such areas at seeding time can be reduced to a minimum by using chemicals instead of the plow, disc, or sweep cultivator in the renovating process. This new, interesting development is discussed in Chapter 5.

The Yearbook of Agriculture: Grass (p. 68) makes these statements:

> Thousands of measurements have been made at the 10 research stations established since 1930 by the Department of Agriculture (Soil Conservation Service). Both soil and water losses have been measured from standardized plots on different types of land used for crops including grass, legumes, trees, and shrubs.
>
> From clean tilled crops, grown at the 10 older stations, the average soil loss over a period ranging from 6 to 11 years amounted to 42.10 tons an acre annually, and the water loss ranged from 2.50 to 40.32 percent of the rainfall. In contrast, from the same kind of land on the same station, the corresponding losses from grass field were only 0.08 of a ton of soil an acre, and from 0.05 to 8.1 percent of the rainfall. Expressed differently, the soil loss from clean tillage was 526 times as much as that from the same kind of land under grass, and the greatest loss of water from the clean cultivated fields exceeded the corresponding water loss from grass by 5 times.
>
> The greatest effectiveness of grass in holding soil and storing water in the ground measured at these stations was at Temple, Texas. There, from black, waxy clay of the Texas blacklands, 20.58 tons of soil an acre (11-year average) were lost where corn was grown—along with 13.6 percent of the rainfall. From Bermuda grass grown on the same kind of land, receiving the same rainfall (an average of 32.7 inches), the soil loss was only 0.02 of a ton an acre; the water loss was 0.05 percent of the precipitation. In other words, soil loss from corn was 1,029 times greater and the water loss 272 times greater than the corresponding losses from a cover of grass.
>
> Many thousands of additional measurements have confirmed the effectiveness of holding back water—storing it in the reservoir of the soil.

Build soil structure with pastures

The mat growth of grasses acts as an insulator, giving the soil a more uniform temperature. This mat also reduces evaporation of soil moisture and, as has alredy been discussed, reduces water run-off. These conditions favor the activity of soil microorganisms. Then too, as the roots die they add valuable organic matter, which in turn increases the beneficial microbial activity in the soil.

One of the most effective ways of building up the organic matter content of the soil is to grow grass on it. The effect of the added organic matter plus the tendency of the root hairs to develop a granular structure brings about a marked improvement in the productive capacity of the soil.

Such soils are much more open and spongy than compact, dense soils. They have larger pore spaces, which encourage ideal water-air relationships. On the other hand, soils low in organic matter have small pore spaces and very often have unfavorable moisture-air relationships.

Lighten soil with pastures

Soil density increases as a result of organic-matter depletion brought about by continuous cultivation. On land that repeatedly was in corn in northern Ohio, it was found that a cubic foot of this soil weighed 81.7 pounds. This same soil, under virgin conditions, weighed 65.5 pounds per cubic foot. The altered pore space was a reduction from 63.3 percent in the virgin soil to 50.5 percent in the cultivated soil. The organic matter content was a reduction from 66 to 44.7 tons per acre. Moreover, the remaining pores were small and less effective as water channels.

At Bethany, Missouri, nine experimental plots on an 8 percent slope had an organic matter content of 3.25 percent in the top 7 inches. During the next 13 years, the plots were cropped in various ways or were permitted to lie fallow. Two of the plots, numbers 6 and 7 (see Table 3.5), were treated with lime and phosphate. In the thirteenth year, tests were made to determine the soil organic matter content. The beneficial effect of grass is vividly illustrated by these data.

Further evidence of how sods conserve and build organic matter

Table 3.5

The Effect of Cropping Systems on Soil Organic Matter[1]

Plot No.	Cropping System	Organic Matter Percent
2	Corn annually	2.23
3	Rotation corn-wheat-clover	3.23
4	Rotation corn-wheat-clover	3.23
5	Rotation corn-wheat-clover	3.15
6	Rotation corn-wheat-clover	3.38
7	Alfalfa, continuous	3.93
8	Bluegrass, continuous	3.61
9	Topsoil, clean fallow	1.93
10	Subsoil, clean fallow	1.41

[1]*Source:* Missouri Agricultural Experiment Station.

is demonstrated by other Missouri data. After six years of cropping to corn and oats, the soil organic matter at plow depth of a Shelby loam area in northern Harrison County averaged 1.3 percent. On an adjoining area that had remained in grass during the same period the corresponding organic matter content was 5.91 percent.

CHAPTER 4

Important Pasture Grasses and Legumes

The livestock farmer profits most from growing a variety of grasses and legumes, either as a mixture or in sequence, that provides grazing as long as possible throughout the growing season in any one area. This means that he or she will rotate the animals from one grass or grass-legume stand to another, thereby gaining maximum use of the growing habits of the different species.

Chapter 4 discusses merits of the more important grasses and legumes and their adaptation to different areas of the United States. Following, in this chapter, is a list of the important grasses and legumes in outline form. This listing includes the approved practices necessary in utilizing these plants as well as the major advantages and disadvantages. It is a desirable practice to plant grasses and legumes best adapted to the conditions found on each farm.

Alfalfa

Advantages

1. It is widely adapted—it can be grown on some soils in every state in the United States.
2. It provides an abundance of nutritious feed.
3. Because it's a legume, alfalfa, when properly inoculated, can obtain most of its nitrogen requirements from the air.
4. Properly managed, the stands are long-lived.
5. It can be grown alone or in a variety of grass mixtures.
6. It is resistant to drought.

Disadvantages

1. Alfalfa does not thrive on acid or non-fertile soils.
2. The seed is relatively expensive.
3. Bloat is a problem when animals graze on pure stands or mixtures high in alfalfa.

Approved practices in using alfalfa

1. A good seedbed is essential.
2. The seed should be inoculated.
3. Seeding rates vary from 4 to 20 pounds per acre.
4. Plant varieties adapted to local conditions. There are many varieties of alfalfa, and the farmer should check with the county agent or state Agricultural Experiment Station for varieties adapted to that state or area.
5. Fall seedings must be made early enough to insure sufficient growth to protect the plant against late winter heaving; otherwise spring seedings are recommended.

Bahiagrass

Advantages

1. Bahiagrass spreads by short, heavy runners and forms dense, tough sod even on droughty, sandy soil.
2. It is adapted to the Southern Coastal Plains area where temperatures are not severe—an area where many other grasses fail.

Disadvantages

1. Bahia will not survive even moderately severe winter temperatures.
2. Frequently, difficulties are encountered in establishing satisfactory stands.
3. Yields are relatively low unless the soil is well fertilized.

Courtesy, USDA

Fig. 4.1—Bahia on sandy soils in the South.

Approved practices in using bahiagrass

1. There are several varieties that are used rather extensively in Florida—Pensacola and Paraguay.
2. To establish, seedings should be made in the spring after corn planting. This means that in the area of adaptation, the seeding should be made in April or May, at the rate of 20 to 25 pounds per acre.
3. As with other grasses, a good seedbed is essential.
4. Lime and fertilizer are needed on most soils adapted to bahiagrass to insure good stands and satisfactory growth.

Bermudagrass

Advantages

1. Bermudagrass grows luxuriantly during the hot summer.

2. It spreads by underground rootstocks called rhizomes and above-ground rooted stems called stolons.

3. When properly fertilized, improved strains of bermudagrass produce tremendous quantities of herbage.

Disadvantages

1. Bermudagrass does not survive extended periods of cold weather.

2. The high-yielding strains, Coastal and Midland, are very sparse seed producers. Therefore, the only practical way to expand either of these strains is to use stolons (above-ground rooted stems).

3. It is difficult to eradicate when no longer wanted.

Approved practices in using bermudagrass

1. Many methods are used to plant stolons of the improved strains of bermudagrass. The common practice is to plow furrows 4 to

Courtesy, USDA

Fig. 4.2—New strains of bermuda are high yielders.

6 feet apart and drop the stolons every 2 feet apart in the furrows. Deep planting is important if the stolons are not watered. This prevents them from drying out.

2. A complete fertilizer such as a 12-12-12 or a 16-16-16 should be applied at the rate of 500 to 600 pounds per acre just ahead of planting. For rapid establishment, nitrogen fertilizer should be applied in mid-summer at 100 to 170 pounds of nitrogen per acre. Continued heavy nitrogen fertilization is necessary every year to insure high yields of protein-rich herbage. Experiments at the Georgia Agricultural Experiment Station have shown that the cheapest herbage is produced when at least 200 pounds of nitrogen (600 pounds of ammonium nitrate) are applied annually.

Big Trefoil
(Marsh, Greater, Major Trefoil)

Advantages

1. Big trefoil is a long-lived, spreading (by rhizomes), perennial legume.
2. It will make more growth than birdsfoot trefoil, where adapted.
3. It is adapted to wet, poorly drained, acid soils.
4. Big trefoil is tolerant of some winter submergence in coastal areas, especially the smooth variety.
5. It is compatible with meadow foxtail, tall fescue, and reed canarygrass in low, wet areas.
6. It provides a palatable, high-protein pasture.
7. Since it has underground rootstocks (rhizomes), it is more resistant to grazing injury than other legumes not so structured.
8. The plant provides forage and winter cover crop where adapted.

Disadvantages

1. Big trefoil requires a special inoculant.
2. It is slow in establishing itself and usually does not provide much forage the first year after seeding.

3. Seeding in old or run-down pastures brings only varying degrees of success.

4. A nurse crop to establish new stands is not recommended since big trefoil is not a good competitor for light and moisture while establishing itself.

5. Its range seems to be restricted to coastal or near coastal areas.

Approved practices in using big trefoil

1. A firm seedbed, free of weeds, is essential.

2. Beaver variety of big trefoil has hairy stems and leaves, while the Columbia variety is smooth-leaved.

3. The smooth variety seems to be preferred by livestock and has more winter submergence tolerance.

4. Seeding is normally done in fall or early spring at the rate of 3 pounds per acre, or 5 pounds per acre if a good stand is needed more quickly.

5. Broadcast seeding, with a drill or by hand, followed by rolling with a corrugated type of roller to cover the seed to a depth of ¼ inch is recommended.

6. Big trefoil fits well into pasture mixture with various grasses.

Birdsfoot Trefoil

Advantages

1. This legume grows well on somewhat poorly drained soils and does not require as much lime as alfalfa.

2. The plants are deep-rooted and make good growth during the summer when more shallow legumes are dormant.

3. The forage is palatable to most classes of livestock.

4. Unlike other legumes, trefoil pastures do not seem to be troubled with a problem of bloat in ruminant animals.

5. Being a legume, birdsfoot trefoil can obtain most of its nitrogen from the air when properly inoculated.

Fig. 4.3—Birdsfoot trefoil is a cool-season legume that does not induce bloating in ruminant animals.

Disadvantages

1. The establishment of good stands is very difficult.
2. Birdsfoot trefoil seedlings grow slowly. It is essential that the companion grasses and legumes be seeded lightly so as not to shade out the young birdsfoot trefoil seedlings.

Approved practices in using birdsfoot trefoil

1. A prepared firm, weed-free seedbed is needed since the seed is very small.
2. In the North, seeding is done in the spring; further South, seedings can be made either in the spring or in the fall.
3. Birdsfoot trefoil is usually seeded in mixtures with one or two grasses.
4. In mixtures, 1 to 4 pounds of seed per acre is usually sufficient to give a satisfactory stand on a firm, well-prepared seedbed.

5. Inoculation is a must. Special birdsfoot trefoil inoculum is available from inoculum supply firms.

Blue Grama Grass
(Example of Native Short Grass)

Advantages

1. Blue grama is a warm-season, perennial bunch grass that forms a sod under heavy grazing.
2. It has the most extensive distribution of all grasses on soils of the plains, mesas, and sandhills.
3. It is often associated with western wheatgrass, buffalograss, and needle grasses.
4. The grass produces a good amount of nutritious forage for summer and winter grazing, where adapted.

Disadvantages

1. It is basically a warm-season grass.
2. Its leaves are mostly basal and may tend to curl.
3. Blue grama seed is not easily obtained.

Approved practices in using blue grama grass

1. Blue grama is generally used in mixtures with other short and mid-grasses.
2. Manage grazing to maintain stands.
3. Reseed as necessary and practical.
4. Apply nitrogen fertilizer in the spring, if practical.

Smooth Bromegrass

Advantages

1. Smooth bromegrass is a long-lived perennial grass.

2. It is adapted especially to regions of moderate rainfall and low-to-moderate summer temperatures.

3. It is one of the best pasture grasses in its area of adaptation.

4. It starts to grow early in the spring and continues when growth is not limited by lack of moisture throughout the summer.

5. Smooth bromegrass is palatable to all classes of livestock.

6. Smooth bromegrass fits well into grass-legume mixtures, especially as a companion crop to alfalfa.

7. Its use in combination with legumes reduces the bloat hazard.

Fig. 4.4—Smooth bromegrass is a cool-season perennial, highly productive and palatable.

Disadvantages

1. Smooth bromegrass requires a fertile soil, one especially high in nitrogen.
2. It cannot stand extended periods of hot, dry weather.
3. It is subject to attack by several plant diseases.

Approved practices in using smooth bromegrass

1. The selection of an adapted strain is a "must." There are two distinct types of bromegrass:
   a. The southern type, best adapted to the Corn Belt states and to sections of the Central Great Plains; and
   b. The northern type, which is generally best adapted to the Northern Great Plains and to Canada, where long periods of dry weather seldom occur.
2. For rapid establishment of a stand, the use of a fertilizer high in nitrogen is recommended.
3. Seed on a well-prepared seedbed using 8 to 15 pounds of seed per acre, the amount depending on the other grass or legumes in the mixture.
4. Use a cultipacker or similar tool to insure good contact between the seed and the soil.

Crested Wheatgrass

Advantages

1. Crested wheatgrass is a hardy perennial bunch-type grass.
2. It is well-adapted to the cool, dry areas of the Northern Great Plains, the Intermountain regions, and the higher elevations in the Rocky Mountain states.
3. It has a root system which is wide-spreading and deep-penetrating.
4. Crested wheatgrass can survive drought and cold, withstand close grazing, and compete well with weeds and associated grasses.
5. Growth starts early in the spring.
6. It is highly palatable and a fairly good yielder.

Disadvantages

1. Crested wheatgrass stops active growth during the hot, dry period of summer.
2. The bunch-type growth is an objection under some conditions.

Approved practices in using crested wheatgrass

1. A well-prepared seedbed aids in establishing a satisfactory stand.

Courtesy, USDA

Fig. 4.5—A growth of crested wheatgrass.

2. Excellent stands have been obtained from solid drilling or broadcasting at the rate of 4 to 8 pounds of clean seed per acre.

Crimson Clover

Advantages

1. Crimson clover is a winter legume.
2. In the deep South, crimson clover furnishes grazing during the winter months. Farther north but still in its area of adaptation, it provides lush, early spring grazing.
3. With the introduction of the hard-seeded varieties such as Dixie and Auburn, the crop behaves like a perennial. The hard seeds fall on the ground but do not germinate until fall.
4. The forage is relished by all classes of livestock.

Disadvantages

1. Crimson clover provides no grazing during the hot summer months.
2. With the exception of the hard-seeded varieties, the crop has to be planted annually.
3. It is adapted only to areas where warm climates prevail.
4. Crimson clover winter-kills in areas of low temperature.

Approved practices in using crimson clover

1. Seedings are usually made in August or September.
2. The recommended rate of seeding is 15 to 25 pounds per acre of inoculated seed.
3. A firm, well-prepared seedbed hastens germination and the establishment of a good stand.
4. It is often seeded in a mixture of ryegrass or small grains for "winter" pasture.

Dallisgrass

Advantages

1. This grass, a relative of bahiagrass, produces well during the summer months.
2. Dallisgrass seldom forms a dense sod, so it is well suited to seeding in combination with legumes or other grasses.
3. Dallisgrass will maintain a good stand and will remain productive for years when adequately fertilized and managed.

Disadvantages

1. Dallisgrass requires a moist soil and grows best only on soils well supplied with organic matter.

Courtesy, USDA

Fig. 4.6—Dallisgrass produces well during the summer months.

2. Dallisgrass demands more fertility than its relative, bahiagrass.

3. The germination of the seed is often low because of the fungus disease "ergot" (*Claviceps paspali*), which attacks and destroys the seed.

4. It is difficult to eradicate.

5. Livestock do not like to eat it when mature.

Approved practices in using dallisgrass

1. Seeding should be made after corn planting time on a well-prepared and adequately limed seedbed.

2. Dallisgrass requires more plant food nutrients for good results than bahiagrass.

Johnsongrass

Advantages

1. Johnsongrass is a vigorous grass that spreads by underground rootstocks.

2. When kept grazed down, johnsongrass makes a tremendous amount of herbage which livestock use to make good gains.

3. It furnishes grazing during the hot, dry summer months.

Disadvantages

1. In many sections, this grass is a pest.

2. It is difficult to destroy. In areas where the land is not infested, it is not recommended for pasture grass.

Approved practices in using johnsongrass

1. The seeding of this grass is not recommended except upon areas already infested with johnsongrass. The Mississippi Agricultural Experiment Station has developed some strains of johnsongrass which do not spread as widely from roots as some of the older strains. The purpose of the Mississippi breeding program is not only to gain the

high yield associated with this grass but also to obtain a strain that can be killed by cultivation.

There are those who feel that with the advent of chemical herbicides to control this grass, farmers may be able to make more use of johnsongrass as a pasture plant. Certainly, any farmer should consult an agricultural specialist before planting this grass on the farm.

Kentucky Bluegrass

Advantages

1. Bluegrass herbage is nutritious, rich in minerals and vitamins.
2. Bluegrass is relished by all classes of livestock.
3. Bluegrass spreads by underground rhizomes and thus maintains a dense stand on suitable soil.
4. It is adapted to soils of the northern states and the mountainous and cooler locations further south.
5. It is an ideal companion for common white clover and is fairly well suited for growing with birdsfoot trefoil.
6. It appears spontaneously in its area of adaptation when soil conditions are favorable.

Disadvantages

1. Although bluegrass furnishes excellent grazing in early spring and again in the fall, it goes dormant during hot, dry periods in the summer and turns brown unless water is applied.
2. To thrive best, bluegrass requires a soil high in phosphorus and lime such as that found in the famed bluegrass region of Kentucky.
3. Yields are consistently low.

Approved practices in using Kentucky bluegrass

1. Most bluegrass pastures are not sown. The grass appears spontaneously in areas where adapted.
2. It can be grazed closely during the spring and fall. The sod is not severely damaged by animals during periods of wet weather.

Bluegrass is losing its popularity since it is a low yielder, especially during the summer months.

Ladino Clover

Advantages

1. This legume is popular with all classes of livestock, including poultry and hogs.
2. Ladino clover is very productive. It has been said that an acre of vigorously growing Ladino clover produces enough feed, if completely harvested, to produce 30,000 pounds of 4 percent milk.

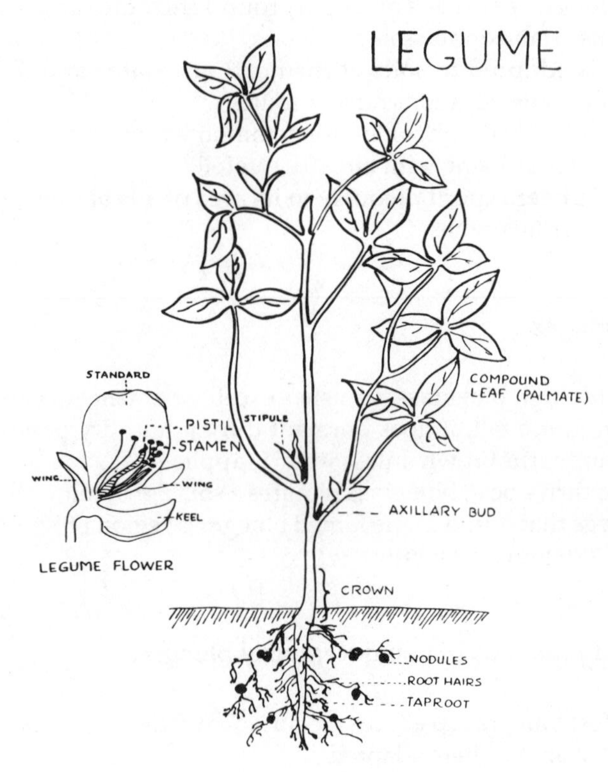

Drawing by C. McVickar

Fig. 4.7—Legume plant structures.

3. The herbage is high in minerals, rich in protein, rich in vitamins, and low in fiber.
4. Ladino clover tolerates wetter soils than alfalfa. As a result, it is injured less by winter heaving.
5. Ladino clover re-establishes itself by natural seeding.
6. It recovers rapidly after being grazed closely.
7. Being a legume, Ladino clover fixes atmospheric nitrogen.
8. Ladino clover is the ideal legume in numerous grass-legume pasture mixtures.
9. Ladino clover can be started by simply tossing the seed on top of the soil in the grass, providing it is then irrigated properly.

Disadvantages

1. If planted too deeply, germination is poor.
2. Ladino clover cannot tolerate long periods of severe drought.

Courtesy, USDA

Fig. 4.8—Ladino clover is widely used in pasture mixtures.

3. The seed of Ladino resembles common white clover seed. The young plants are identical in appearance. Thus a reliable seed source is the farmer's only guarantee of getting Ladino instead of common white.

4. Bloat can be a serious problem on pastures.

5. Close and continuous grazing will kill out the stand. A common practice is to divide large fields into smaller areas. The animals are turned on one area and, when it is grazed down, rotated to another. Under this system, Ladino clover recovers quickly and is ready for grazing again in two or three weeks.

6. Ladino clover must be grazed lightly during the fall months. Unless the plants make sufficient growth to build up good root reserves, winter killing may be severe. On the other hand, stands have been occasionally destroyed by smothering when heavy growths are permitted to go into the winter.

7. Ladino clover may have a laxative effect on the animals, and they may lose weight for a short time before becoming accustomed to it. This difficulty is largely overcome if the animals are provided with dry hay at all times and well-fed before being turned into Ladino clover pastures. Permitting them to graze for only short intervals during the first four or five days will also help.

Approved practices in using Ladino clover

1. A firm, well-prepared seedbed is essential.

2. The seed should not be covered more than ¼ inch. The soil should be firmed by either cultipacker or press wheel immediately after seeding Ladino clover.

3. The seed should be properly inoculated with a suitable inoculant.

4. In the South, seedings are usually made in the fall to avoid severe weed competition. In the North, spring plantings are almost a necessity to avoid winter killing of the young seedlings.

5. Seeding at the rate of ½ pound to 2 pounds per acre gives good stands with grasses and other legumes when planted on well-prepared seedbeds.

6. The bloat hazard can be partially eliminated by giving animals access to dry hay, straw, or other grassy pastures, and ample salt and water. A good proportion of grass in the mixture also reduces the bloat hazard. Cattle and sheep should not be turned into Ladino for

Courtesy, USDA

Fig. 4.9—Ladino clover–grass mixtures are widely adapted.

the first time when the clover is wet or when the animals are very hungry. The danger of bloat is materially reduced if the animals are well-fed before grazing vigorously growing Ladino clover.

Lespedeza

Advantages

1. Lespedeza is especially suitable for use on pasture soils of low fertility.
2. Good stands are easy to obtain.
3. Seed prices are relatively low. Establishing a stand can be done economically.
4. Lespedeza yields considerable forage when grass companions are more or less dormant.
5. Lespedeza is a legume and, therefore, tends to fix nitrogen.

Disadvantages

1. Lespedeza is not as rich in minerals as most other legumes.
2. The best stands often do not develop until the third year after seeding.
3. Productive growth is limited to warm, mild summer months.

Approved practices in using lespedeza

1. Lespedeza seed is almost universally included in all mixtures for the southeastern and south-central states. The usual recommendation is 20 to 30 pounds of seed per acre, either broadcast or drilled.
2. Although lespedeza grows fairly well on soils of low fertility, yields can be stepped up materially by applications of complete fertilizer high in phosphate and potash.

Little Bluestem (Example of Native Mid-grass)

Advantages

1. Little bluestem is a bunch grass with occasional short rhizomes and is a perennial.
2. It is a warm-season mid-grass.
3. It is adapted to Northern Plains, dry hills, and open woods of Inter-mountain areas.
4. The grass performs well on medium to sandy and gravelly soils.
5. It can be used for grazing or hay and has some salt (alkaline) tolerance.

Disadvantages

1. It is not readily eaten by livestock after seedheads are produced.

2. In heavily grazed or continuously grazed areas, little bluestem tends to be replaced by blue grama and other short grasses.
3. It has a comparatively high moisture requirement for good production.
4. The pedicels are covered by short white hairs, rendering forage less palatable.
5. Forage is produced during the warm season only.
6. The leaves are rather rough and narrow.

Approved practices in using little bluestem

1. Seedbeds should be clean and firm since seedlings are small and grow slowly at first.
2. Where wind and blowing soil might damage seedling plants, it may be best to plant in a dead stubble of grass growing there the year before.
3. Fertilizer should be offered for rapid growth.
4. Early fall seeding is more apt to succeed.
5. Planting rates are expressed in terms of pure live seed per acre.
6. New stands should not be grazed until they are well established.
7. Grazing should be managed so the plants remain thrifty.

Meadow Foxtail

Advantages

1. It is a long-lived perennial grass.
2. Growth starts early in spring, making meadow foxtail about the earliest of all cultivated grasses.
3. The dense rooting system reaches a depth of 2 feet under good conditions.
4. An abundance of dark green leaves is produced.
5. The seed germinates very rapidly.
6. Good growth is made in shade and in wet areas.
7. Meadow foxtail grows best in moist, fertile soils.
8. It makes a leafy, palatable hay, high in protein.

Disadvantages

1. Seeding in grain drills is difficult since the kernel is enclosed in a fluffy, membranous covering.
2. Meadow foxtail responds readily to nitrogen fertilizer, and older stands do not produce satisfactory yields of forage without added nitrogen.
3. It matures unevenly, which can lead to considerable shattering of seed.
4. Seed is hard to clean and is light in weight.
5. Meadow foxtail should not be considered superior to such grasses as timothy, orchardgrass, and bromegrass where these grasses are adapted.

Approved practices in using meadow foxtail

1. The seedbed should be fine, firm, and free of weeds.
2. Seeding should be done at the rate of 10 to 14 pounds of good quality seed per acre on a well-prepared seedbed, where moisture is available for maximum germination and continuous growth.
3. When seeded in combination with other grasses and legumes, 3 to 5 pounds per acre are generally used.
4. Seed is generally placed in the grain box of a grain drill and must be stirred frequently to prevent matting due to the fluffy covering on the seeds, or meadow foxtail can be broadcast seeded.
5. Drilled seed should be covered no more than 1 inch.
6. Meadow foxtail can be seeded by grain drill in renovated (disced) shallow sods of rough meadowlands.
7. It produces a succulent forage during spring, early summer, late summer, and early fall where adapted.

Orchardgrass

Advantages

1. Orchardgrass is one of the most widely adapted pasture grasses.
2. Its longevity, leafiness, and ability to withstand adverse soil and climatic conditions in a humid area make it a very desirable pasture grass.

3. Although orchardgrass thrives best on a rich soil, it also grows relatively well on light soils of medium fertility and moist, heavy soils.

4. Orchardgrass is one of the best grasses for production in shady areas.

5. Orchardgrass is relatively cold-resistant and continues to grow until the first severe frost.

6. New growth starts early in the spring.

7. Neither timothy nor bromegrass equals orchardgrass in midsummer production of leafy forage. It is more tolerant to heat, drought, and low fertility than bromegrass.

8. It is well suited for use with Ladino clover and lespedeza in pasture mixtures.

Fig. 4.10—Orchardgrass, popular in pasture mixtures, is a heavy producer.

Disadvantages

1. The bunch-type growth is an objectionable feature in certain pastures. The herbage is not as nutritious as that of bromegrass or timothy, especially when growth is large and heavy.

2. Careful management is required to maintain the legume or legumes in the stand with orchardgrass. However, if early orchardgrass growth is removed by grazing or mowing, legumes, especially Ladino clover, will not be crowded out of the combination, and the mixture will continue to produce abundant, high-quality grazing during the summer when many pasture crops are short or dormant.

Approved practices in using orchardgrass

1. Orchardgrass can be sown alone or in combination with legumes and other grasses.

2. Both fall and spring seedings are successful in milder regions. Spring plantings are recommended for colder sections.

3. Like other grasses, a firm seedbed is desirable.

4. The growth of pure stands of orchardgrass can be greatly stimulated by heavy application of nitrogen fertilizer. Many farmers fertilize heavily to stimulate early spring growth or to extend the grazing season later into the fall.

5. Seed at the rate of 6 to 15 pounds per acre alone or 3 to 5 pounds per acre with alfalfa.

Range Clovers
(Bur, Crimson, Rose, and Subterranean)

Advantages

1. Range clovers grow in most California climates and soils.

2. They do well on brush burns, annual type ranges, and abandoned grain land.

3. Such clovers provide plentiful high-value feed in the spring and make good dry feed in the summer and fall.

4. They are sturdy, widely adapted, prolific seed producers.

5. Rose clover is the most widely adapted range legume for California and will grow on soils where many other forage plants will not survive.

6. Crimson clover is a winter annual legume and a heavy producer on much rangeland.

7. Bur clover is adapted to much of the range in California and volunteers widely.

8. Sub clover is a low-growing winter annual which perpetuates itself by burying its seed heads in the soil and thus is tolerant of close, heavy grazing.

Disadvantages

1. Range clovers are cool-season plants, so they do not produce much forage during warm months.

2. Seed may be hard to come by if these annual plants are not managed to permit their reseeding of themselves.

3. Quite often applications of phosphorus and sulfur are necessary for good range production.

Approved practices in using range clovers

1. October is the optimum month for seeding clovers over most of the area to which they are adapted.

2. Inoculation of the clovers with nitrogen-fixing bacteria is good insurance for stand establishment.

3. Seed treatment with a fungicide may be profitable where seedling diseases are a problem in obtaining stands.

4. Clovers should not be seeded more than ¾ inch deep, using a grain drill having a fertilizer attachment and a grass seeder box with an inexpensive modification.

5. Seedbed preparation is desirable with plowing or discing filling the bill, dependent upon tentative competition of weeds and forbs with the clovers.

6. Sudangrass has been used as a cleanup crop prior to planting clovers.

7. Controlled burning of tall, dense brush provides a desirable seedbed for clovers.

8. Timely grazing is essential to insure permanent stands of clovers.

9. Apply superphosphate, which contains sulfur in addition to phosphorus, in adequate amounts by banding at seeding time and every two to three years thereafter according to needs as shown by soil tests.

Red Clover

Advantages

1. Red clover provides a high yield of nutritious forage the first year after seeding.
2. It is a legume and thus can fix atmospheric nitrogen.
3. This legume grows better than alfalfa or sweet clover on soils that are slightly acid or not too well drained.
4. It is more winter-hardy than alfalfa.

Fig. 4.11—Red clover is a short-lived perennial legume, high-yielding, highly palatable, and nutritious.

5. The red clover plants start off faster than small-seeded clovers such as common white or Ladino.

Disadvantages

1. Most red clovers are short-lived due to diseases and act like biennials. Thus they are short-lived in pasture mixtures.

2. Red clover roots are more shallow than alfalfa roots. Therefore, production is cut during drought periods.

Approved practices in using red clover

1. A small amount, 2 to 8 pounds of red clover per acre in the pasture mixture, is recommended simply to provide a quick growth of forage. Red clover is almost always seeded in a mixture of other grasses and legumes which live longer and take over as the red clover disappears.

2. The seed should be inoculated. A well-prepared seedbed is mandatory. Fall seedings can be made in the South. Spring seedings are recommended in the North to avoid winter killing of the young plants.

3. Liberal amounts of phosphate and potash increase yields materially.

Redtop

Advantages

1. Redtop is a long-lived, sod-forming, easily established perennial grass.

2. It is one of the most widely adapted grasses.

3. It can be used in pasture, hay, and lawn mixtures.

4. Redtop has an acidity tolerance range of from 5.0 to 7.0 and performs well on wet soils.

5. Its seed retains viability for long periods.

6. If well-managed, redtop is relished by most classes of livestock in pasture and hay mixtures.

7. It grows well on level clay-pan soils and infertile soils.

Disadvantages

1. Unless used in mixtures with legumes, redtop requires considerable nitrogen fertilizer for good growth.
2. It is less productive and less palatable than orchardgrass.
3. It does not survive drought any better than orchardgrass.
4. Redtop tends to develop a "bitter taste" if growth becomes too mature.
5. The quantity and quality of nutrients are less than from many other grasses.

Approved practices in using redtop

1. A fine, firm seedbed is preferred but redtop will germinate under most conditions unless seeded too deeply.
2. Generally it is used in mixtures with other grasses to establish vegetative cover quickly.
3. It can be seeded in early spring, early fall, or late fall.
4. When used for hay, redtop should be harvested before or during bloom.
5. It is adapted for use in grassed watercourse seedings.
6. It gives fairly good growth, especially for early spring and fall pasturing.

Reed Canarygrass

Advantages

1. Reed canarygrass is a long-lived, sod-forming, leafy perennial.
2. It is flood-tolerant and high-yielding.
3. A large part of its growth occurs during the summer, when additional pasture often is needed.
4. The grass tends to grow in dense bunches and spreads underground by short, creeping rootstocks.
5. Its best growth is on fertile, moist, swampy soils.
6. It is one of the most drought-resistant grasses.
7. It is palatable to most classes of livestock (when grazed before seedhead development).

8. Reed canarygrass is adapted to most muck areas in the northern half of the United States.

Disadvantages

1. Light-colored soils must be fertilized heavily with nitrogen or manured heavily for successful production of this grass.
2. Reed canarygrass germinates slowly, and spring seedings do not compete well with summer annual weeds.
3. The seed is generally high-priced.

Fig. 4.12—Reed canarygrass is a long-lived perennial adapted to wet or dry soils.

4. Only alsike clover is adapted to wet areas where reed canarygrass is well suited.
5. Seed shatters easily, especially in dry, warm, windy weather.
6. The grass becomes coarse and unpalatable unless grazing is managed carefully.

Approved practices in using reed canarygrass

1. Select an adapted, high-yielding, palatable variety.
2. A fine, smooth, well-packed seedbed is preferred where possible.
3. Prepare seedbed in the dry part of summer in poorly drained areas to be seeded in late summer or early fall.
4. Reed canarygrass is often seeded in a mixture with timothy.
5. Stands may be obtained by discing in sections of rootstocks from established reed canarygrass areas.
6. It should be grazed fairly heavily to maintain palatability.

Ryegrass

Advantages

1. Ryegrass stands are easy to establish.
2. Ryegrass develops rapidly after seeding—usually to the detriment of the better species.
3. When sown in combination with winter grains for temporary pasture, it makes a desirable understory grass and increases the length of the grazing season.
4. Seed is relatively cheap.

Disadvantages

1. Generally ryegrass is short-lived.
2. The forage becomes coarse and stemmy in mid-summer without irrigation.
3. Ryegrass may offer severe competition to the slower-starting grasses and legumes in the mixture.

Approved practices in using ryegrass

1. Ryegrass seedings can be made in the fall or early spring. Fall seedings are recommended in the more temperate regions; spring seedings are best where winters are severe.
2. The seed should be sown approximately ½ inch deep.
3. When sown alone for grazing, seeding at the rate of 20 to 25 pounds of seed per acre is recommended. When sown with a small grain or legume for annual pasture, 8 to 10 pounds per acre will give a satisfactory stand.

Small Grains

Advantages

1. Small grains sown alone or in combinations can provide early spring grazing, late fall grazing, and winter grazing in areas where mild temperatures prevail.
2. Stands are easily established.

Disadvantages

1. Being annuals, small grains must be planted every year. Ground preparation is costly.
2. Annual soil preparation favors soil erosion.

Approved practices in using small grains

1. Small grains seeded for pasture are normally seeded at twice the planting rate for grain and a few weeks ahead of the seeding date for grain production. This early planting hastens the grazing. The seeding is made with a grain drill. Liberal amounts of fertilizer, especially nitrogen, are used to stimulate rapid growth.
2. Many farmers use a combination of small grains so as to capitalize on variable soil and climatic conditions.
3. Often ryegrass and crimson clover are added to the small grains to extend the grazing period.

4. Inclusion of vetch or other adapted annual legumes produces forage of higher protein.

5. Oats, properly managed, can be used for hay, grazing, silage, or grain.

6. Rye, properly managed, can be used for grazing and grain, but does not make good silage.

7. Winter wheat that has been planted for grain harvest should not be grazed after about May 1.

Sudangrass and Hybrids

Advantages

1. Sudangrass and its hybrids are high-producing annual grasses adapted to all areas of the United States.

2. Sudangrass and hybrids provide abundant grazing during the warm summer months, when perennial species are normally low producers.

Fig. 4.13—Sorghum-sudangrass hybrids may be harvested for hay (foreground) or silage (background).

3. They can be used as warm-season emergency or supplemental pasture, hay, greenchop, or silage, especially on dairy farms.

4. Sudangrass and hybrids do best on deep, moderately to highly fertile soils where the growing season has a relatively long warm period, but will grow on a wide range of soils, if nitrogen is supplied.

5. They are equal to the best sorghums in enduring drought, although shortage of moisture lowers yields.

6. They perform well under irrigation and usually crowd out weedy grass plants.

7. Sudangrass and hybrids can be used to "condition" choice rangeland prior to seeding improved pasture species.

Disadvantages

1. Since they are annuals, sudangrass and hybrids need to be established every year.

2. They require a relatively warm growing season for best results.

3. They may be injured or killed by light frosts.

4. They are heavy users of nitrogen, which must be supplied from the soil or fertilizer application.

5. They can cause prussic acid poisoning in livestock when grazed indiscriminately after prolonged dry weather, too soon after clipping, or after a frost.

6. Sudangrass and hybrids are difficult to cure as hay because of their heavy, juicy stems, and they produce lower yields of silage per acre than corn or taller-growing forage sorghums.

Approved practices in using sudangrass and hybrids

1. A firm seedbed insures even germination, which is dependent on enough moisture in the soil.

2. Sudangrass and hybrids may be seeded from mid-April to July in the southern half of the United States and two weeks after corn planting time in the northern United States.

3. Most of the annual weeds can be eliminated during seedbed preparation by application of recommendation rates of suitable preemergence herbicides.

4. A grain drill with press wheels works well for seeding. In semiarid areas, every other hole of the grain drill may be stopped up.

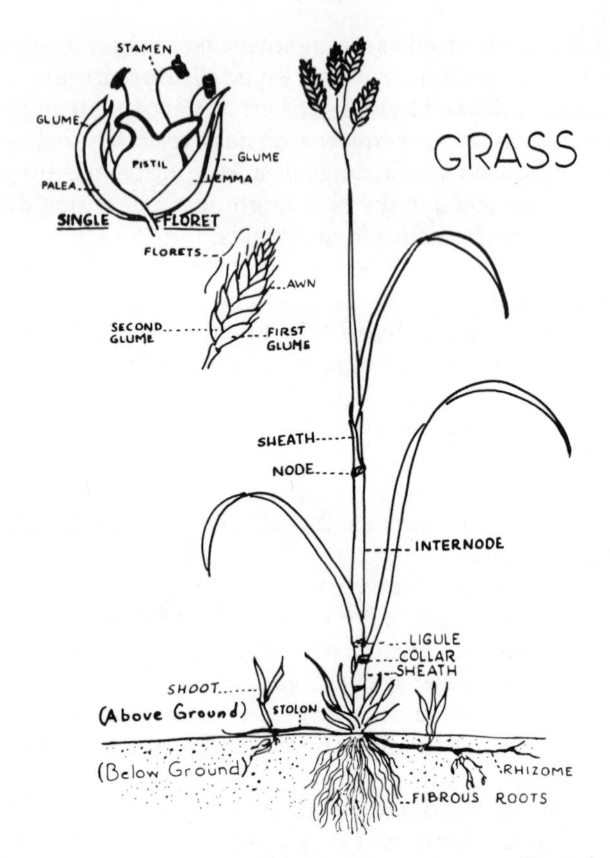

Drawing by C. McVickar

Fig. 4.14—Schematic drawing of the morphology of a grass plant.

5. Grazing of sudangrass should be delayed until it has attained a height of 15 to 18 inches. Sorghum-sudangrass hybrids should be 24 inches tall.

6. Normally a rate of 25 to 35 pounds of seed per acre is recommended in humid areas, 12 to 15 pounds in dry areas if close drilled or broadcast, and 3 to 6 pounds for rows. Under irrigation, 15 to 20 pounds of seed gives good stands, since excellent germination can be expected.

7. There is little danger of cattle being poisoned by sudangrass if the following precautions are taken:

a. Seed containing even a trace of sorghum or johnsongrass should not be sown.

b. Cattle and sheep should not be turned on sudangrass before it has grown to 15 inches.

c. Sudangrass that has been overgrazed or had its growth checked by drought or frost should not be grazed until the new shoots are 15 to 18 inches high.

d. If you suspect prussic acid poisoning, call your veterinarian promptly.

Sweet Clover

Advantages

1. Sweet clover grows vigorously in regions of the United States having 17 inches or more rainfall annually, provided the soil is well fortified with calcium and other plant food nutrients.
2. It produces a volume of feed when many pasture plants are dormant.
3. Being a legume, sweet clover, when properly inoculated, can obtain most of its nitrogen from the air.

Disadvantages

1. The herbage is not relished by animals when they are first turned on sweet clover.
2. Seedings have to be grazed heavily and with care to avoid woodiness.
3. Livestock bloat can be a serious hazard.
4. Sweet clover is adapted only to well-limed soils rich in phosphorus and potassium.

Approved practices in using sweet clover

1. Stands of sweet clover are easily obtained on soils well suited to the growing of this legume.
2. Rate of seeding varies from 8 to 20 pounds per acre.
3. Seedings can be made in a small grain crop with good success.
4. Inoculation of seed is necessary unless sweet clover has been grown on the same land a few years previously.

Tall Fescue

Advantages

1. Tall fescue is adapted to a variety of soil and climatic conditions.
2. It is a top producer and can be grazed closely.
3. It combines well with Ladino clover, white clover, or red clover.
4. Tall fescue is more resistant to diseases and drought than orchardgrass.

Fig. 4.15—Tall fescue is a good grass for cool-season pastures.

5. In milder climates, fescue stays green throughout the winter.
6. It is easy to establish a satisfactory stand.
7. It grows satisfactorily on both well and poorly drained soils.

Disadvantages

1. The palatability of tall fescue is low during the warmer parts of the year. An endophytic fungus is suspected to contribute to lower palatability and animal utilization.

2. Careful management is required to keep this vigorously growing grass from getting the upper hand over the legumes in the mixture. However, when the fescue is kept under control by either grazing or clipping, satisfactory stands of legumes can be maintained with tall fescue.

Approved practices in using tall fescue

1. Seedings can be made either in the fall or in the spring. Spring seedings are preferred where winters are severe.

2. A firm seedbed helps to obtain good seed germination.

3. The amount of seed usually recommended is 10 to 15 pounds per acre.

4. The judicious use of nitrogen to keep the fescue growing vigorously apparently increases the palatability of the forage.

5. The fescue should be kept grazed or mowed down when grown in combination with legumes to prevent the grass from shading out the lower-growing clovers.

Tall Oatgrass

Advantages

1. Tall oatgrass is a hardy, upright-growing perennial.

2. Tall oatgrass comes on early in the spring and remains green until late in the fall.

3. Although it does not form a good sod, tall oatgrass seems to stand pasturing well and furnishes abundant grazing.

4. It prefers well drained soils and seems to be especially adapted to light sandy and gravelly soils.

Disadvantages

1. Tall oatgrass does not propagate by rhizomes and therefore tends to be bunchy.
2. This grass does not do well under shade.
3. The poor seed habit is a drawback. It produces an abundance of seed, but harvesting is difficult because the seed shatters before it fully matures. Also, the seed is often of low viability.

Approved practices in using tall oatgrass

1. The usual rate of seeding is 30 to 40 pounds per acre.
2. Broadcasting is recommended because the coarse, fluffy seed does not feed uniformly through a drill.
3. A well-fitted seedbed is desirable. After the seed is broadcast it should be covered with a cultipacker or harrow.
4. In sections where there is a reasonable amount of moisture in late summer or in the fall, and where winters are not too severe, best stands can be obtained by seeding in September or early October; otherwise, spring seedings are recommended.
5. Best results are obtained when tall oatgrass is pastured in rotation. This gives the plants time for recovery after being grazed down.

Timothy

Advantages

1. Timothy is well-adapted to the cool, humid regions of the Northeast, the Lake States, the Corn Belt, the valleys of the Rocky Mountains, the coastal regions of the Pacific Northwest, and Alaska.
2. It is an ideal companion grass for birdsfoot trefoil. The upright and not excessive growth of timothy does not offer severe competition to the birdsfoot trefoil, which is slow to establish itself.
3. Good seedings are easy to obtain.

4. Seed is relatively inexpensive.
5. It is palatable to all classes of livestock.

Disadvantages

1. Timothy is not as high-yielding as orchardgrass, smooth bromegrass, or tall fescue.
2. Yields are usually less during drought periods.

Fig. 4.16—Timothy is a highly palatable, high-quality grass for cooler regions of the United States.

Approved practices in using timothy

1. Fall seedings are preferred when timothy is sown alone or with fall-planted small grains. Seedings made at this time are less likely to be injured by dry weather in late spring or early summer than seedlings from spring plantings.

2. From 3 to 5 pounds of seed are recommended for fall plantings; 4 to 10 pounds are recommended for spring plantings.

3. Timothy is used extensively in rotated pastures. It goes well in red or alsike clover and is suited for both hay and pasture.

Switchgrass
(Example of Native Tall Grass)

Advantages

1. Switchgrass is a bunch-like perennial that tends to form a sod, especially when heavily grazed.

2. There is a scattered distribution in the Plains States on soils that are medium textured to sandy.

3. The grass produces large amounts of forage that are pastured in late spring and early summer.

4. It has numerous scaly, strong rhizomes.

Disadvantages

1. Switchgrass is a warm-season producer only.

2. The leaves and especially the culms become rank and tough as the plant matures.

3. It is associated with bluestems, sideoats grama, Indiangrass, and others with which it may not compete too well.

Approved practices in using switchgrass

1. Manage grazing to maintain the stand.

2. Switchgrass is best adapted to open ground along drainages or bottoms.

3. Since it is a warm-season grass, it should be used in native grass mixtures, or alternative grazing areas should be provided.
4. It responds to application of fertilizer, especially nitrogen.

White Clover

Advantages

1. White clover is an excellent pasture legume that combines well with orchardgrass, bluegrass, and tall fescue.
2. Its forage is highly nutritious.
3. White clover grows on a wide variety of soils.
4. It re-establishes itself by natural seeding.
5. White clover will stand closer grazing than its "big sister," Ladino clover.

Disadvantages

1. White clover yields less than Ladino clover.
2. It, like Ladino, is sensitive to periods of drought.
3. White clover can cause bloat in livestock.
4. The small seed makes the establishment of a good stand somewhat difficult.

Approved practices in using white clover

1. A well-prepared, firm seedbed is essential because of the small seed size.
2. The seed should be inoculated unless white or Ladino clover has been grown on the field previously.
3. Fall seedings are satisfactory in the South; spring seedings are recommended in the North to avoid winter killing of the young seedlings.
4. The usual recommendation is ½ to 2 pounds of seed in the pasture mixture.

CHAPTER 5

The New Pasture

A good stand is the major requirement of any profitable pasture program. In most sections of the country, stands of vigorous-growing grasses and legumes must be established by the farmer. There are instances where the native vegetation can be converted into profitable pasture by the wise use of lime, fertilizer, and other approved cultural and management practices. Pasture improvement is applicable on land which can carry a type of vegetation which has the inherent capacity to produce large quantities of nutritious herbage. Where such species of plants are not already present, new seedings are required to make the pasture productive. Approved practices for liming and fertilizing and managing an established sod are identical to those discussed in Chapter 7, "Keeping Pastures Productive and Healthy."

Activities Which Involve Approved Practices

1. Getting a good seedbed.
2. Renovating with herbicides.
3. Using lime, sulfur, and fertilizer.
4. Determining lime needs by soil test.
5. Recommending fertilizer requirements.
6. Keeping legumes in the pasture.
7. Determining deficiencies in legumes.
8. Using rotational grazing.
9. Supplying grasses with nitrogen.
10. Boosting yield of grasses.
11. Placing fertilizer so it can be reached by plants.
12. Fertilizing for no-till seeding.
13. Selecting a high-yielding adapted mixture.
14. Planting in spring and late summer.

Fig. 5.1—A good grass cover is a good soil conservation practice.

15. Using a companion crop.
16. Obtaining quality seed.
17. Inoculating seed.
18. Preparing the seedbed.
19. Planting seed.
20. Managing the first year.
21. Increasing legumes in grass pastures.

1. Getting a Good Seedbed

Destroy old vegetation

There is no substitute for a well-prepared seedbed. Generally, seeds of the improved grasses and legumes are very tiny and cannot compete effectively with weeds and other wild plants. If the pasture is being established on land that has not been clean-cultivated it is therefore essential that any old vegetation be thoroughly destroyed before the new seeding is made. Some farmers plow the land; this is a satisfactory method. If the old vegetation is turned under, it usually is

Fig. 5.2—The tillage method of pasture renovation requires nearly complete kill of old vegetation.

completely killed. On steep or rolling land, however, plowing leaves the soil unprotected, and heavy soil losses may occur during periods of rainy weather. Some farmers use heavy disc or tractor tillage-bar implements to rip up the old vegetation. These machines simply uproot the old vegetation and leave it on or near the soil surface. This method of land preparation gives a good kill of the old vegetation when thoroughly done, especially if a dry period follows the operation. On the other hand, if rain falls soon after the land is stirred, much of the vegetation takes root and starts to grow again.

Use trash method for erosion control

This so-called method, however, has the advantage of reducing soil losses due to erosion. The trash on or near the surface counteracts the pounding action of the raindrops and offers mechanical resistance to water run-off. The trash also acts as a mulch, reducing soil crusting and maintaining more uniform soil temperatures, thereby contributing to better conditions for seed germination.

2. Renovating with Herbicides

Early in the 1960's, Dr. Milton Sprague at the New Jersey Agricultural Experiment Station demonstrated the feasibility of using herbicides with moderate discing to prepare a seedbed. This method was especially suitable for renovating pastures on lands subject to severe erosion, when compared to conventional methods.

Sprague's work led to the use of herbicides and no-till seeders for pasture renovation. This method has proven very useful and desirable on steep and/or rocky pastures, especially where soil erosion would be a serious hazard if clean tillage were practiced.

No-till pasture renovation to interseed legumes into pastures is most successful if the pasture is overgrazed sufficiently to reduce the vigor of the grasses and other herbage in the pasture. Lime, phosphorus, potassium, and other nutrients as needed should be applied according to a soil test. A herbicide should be applied which either subdues or kills the existing vegetation. There is a choice of herbicides. Consult your County Agricultural Agent or your State Agricultural Experiment Station for current suggestions. Herbicides that just sub-

Fig. 5.3—No-till seeders place seeds into the soil for good seed-soil contact without tearing up the sod.

due the existing pasture vegetation are useful where the existing grasses are desirable and legumes are interseeded to increase dry matter yield, protein, and mineral content and to supply nitrogen for the grasses.

Herbicides that kill grasses and other vegetation are useful when the grasses in the pasture are to be replaced by a different species or variety. Seed with a no-till seeder which is equipped with a coulter to cut through trash, which prepares a narrow slot in the soil, and which places the seed in this slot to achieve good soil-seed contact. Do not graze until the legumes are well established except when excessive grass recovery following the herbicide application threatens to overpower the seedling legumes.

Alfalfa and red clover have been the more successful legumes established by no-till seeding methods in the humid region of the United States.

3. Using Lime, Sulfur, and Fertilizer

Many of our soils are low in lime and plant-food nutrients. Unless these materials are available in adequate quantities, the new seeding will fail completely or will get off to a poor start and never yield satisfactorily. Fortunately, it is easy to find out if a soil needs lime. All that is required is a representative soil sample which can be easily analyzed for pH (a measure of soil acidity) and active calcium.

Test soil

There is no trick to taking representative soil samples. The Fertilizer Institute illustrated the 10 steps in taking representative samples by a series of cartoon drawings. See Fig. 5.4.

Table 5.1 gives the pH requirements for many of the popular grasses and legumes grown in this country.

4. Determining Lime Needs by Soil Test

If the soil test shows the land needs lime, it is advisable to apply adequate quantities to bring the soil up to the desired pH. If the land is on the alkaline side, it may be possible to lower soil pH by applying gypsum or sulfur. More soils need lime than need sulfur.

get information sheet and soil cartons

divide farm into fields for sampling

take composite sample from each area

don't sample unusual areas

use proper sampling tools

STEPS IN TAKING GOOD SOIL SAMPLES

sample to plow depth

remove soil ribbon from center of slice

mix well in clean pail

fill out information sheet

number samples—keep record

Courtesy, The Fertilizer Institute

Fig. 5.4—Follow these simple steps to take good soil samples.

Table 5.1

The Acidity Scale for Many Grasses and Legumes[1]

pH 4.5 (Very Acid)	pH 5.5 (Medium Acid)	pH 7.0 (Neutral)	pH 8.5 (Very Alkaline)
Such grasses as poverty and broomsedge	Orchardgrass Fescues Bluegrass White clover Birdsfoot trefoil	Alfalfa Sweet clover Bromegrass Crested wheat-grass	Not desirable range for most grasses or legumes

[1]As shown here, pH is a measure of soil acidity and active calcium.

Table 5.2

Approximate Amounts of Different Liming Materials Needed to Raise the pH Value One Unit on Various Soils[1]

Soil	Ground Lime-stone Marl or Oyster Shells	Burnt Lime	Hydrated Lime
	(lbs./acre)		
Light sandy	1,500	840	1,110
Sandy loams	2,000	1,120	1,480
Loams	3,000	1,680	2,220
Silt loams & clay loams	3,500	1,960	2,590

For soils low in organic matter, reduce the above amounts 25%. For soils high in organic matter, increase 100%.

[1]*Source:* Virginia Experiment Station.

Virginia Agricultural Experiment Station Bulletin 136 gives the approximate amount of different liming materials needed to raise the pH value one unit on various soils. This information is given in Table 5.2.

Lower soil to neutral point

When a soil has an unusually high pH index it is often advisable to lower it to or slightly below the neutral point. The amount of sulfur or gypsum needed to accomplish this is given in Table 5.3.

Table 5.3

Pounds of Soil Sulfur (95% Sulfur) Required per Acre to Lower pH Value of Soil to 6.5[1]

Soil pH	Broadcast Application on Whole Area		Band Application	
	Sandy Soils	Clay Soils	Sandy Soils	Clay Soils
7.5	500	900	225	400
8.0	1,250	1,750	400	700
8.5	1,750	1,750 up	700	800 up
9.0	2,500		800 up	

[1]Equivalent amount of gypsum five times sulfur figures.

Soil tests also provide a means of determining what the soil's plant-food inventory may be. This information is essential in making sound fertilizer recommendations for the new seedings.

5. Recommending Fertilizer Requirements

The fertilizer recommendations will, of course, vary depending upon the results of the soil test, soil type, previous treatments, and kind of pasture plants being seeded.

No general fertilizer recommendations can be given that apply broadly across the country. To give the student an insight as to quantities that may be needed, the following quotation is taken from Pennsylvania Agricultural Experiment Station Bulletin 554:

> For most soils of average level of available phosphate, 30 to 60 pounds of phosphoric acid per acre per year usually are recommended for maintenance of high yield. This is the amount of phosphoric acid in 150 to 300 pounds of 20 percent superphosphate. Soils that have not been fertilized recently should receive an initial application of 120 to 160 pounds of phosphoric acid per acre followed by 30 to 60 pounds per year for maintenance.

Grazing animals remove about 12 pounds of phosphate (P_2O_5) per ton of dry matter grazed off.

Potassium is a very important nutrient for pastures. Many soils in the United States are so well supplied with potassium that none needs to be applied. Other soils, however, are not capable of pro-

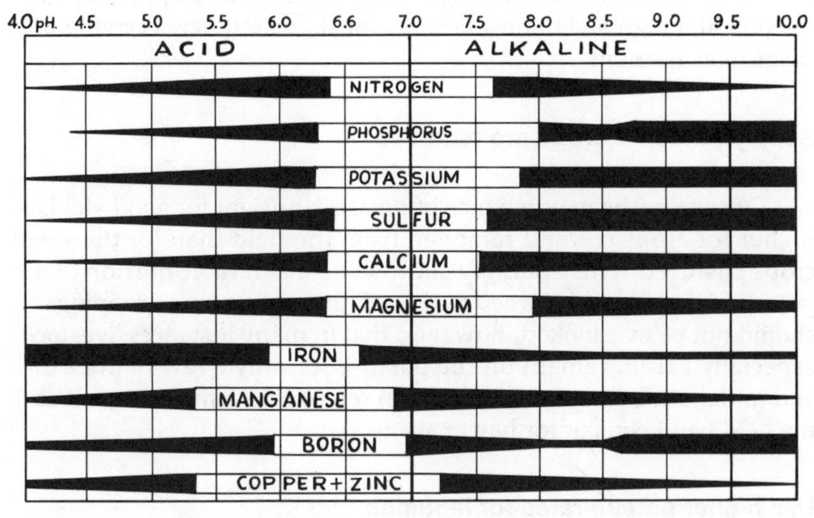

Courtesy, Western Illinois University

Fig. 5.5—Availability of plant nutrient elements as affected by soil reaction pH.

ducing top-yielding pastures without fertilization. Pastures remove 50-60 pounds of potash (K_2O) per ton of dry matter removed by grazing.

Apply yearly

The Pennsylvania bulletin continues:

> It appears to make little difference for maintenance application whether the phosphorus is applied annually or every two or three years provided the same total amount is applied. In most cases, however, the phosphate is applied in combination with potash, and since potash should be applied every year, it may be more practical to apply phosphate at the same time.

6. Keeping Legumes in the Pasture

Potash fertilization and rotational grazing are the "keys" for keeping legumes in pasture mixtures. Although most of our soils are very high in total potash, which slowly weathers to available form, very few of even our better soils release enough potash for top yields of nutritious pasture mixtures year after year. Therefore, to keep production at a high level, most farmers must use fertilizer carrying this plant-food nutrient.

Supply potash if crops are removed

Of course, the maintenance of potash sufficient for good yields is higher for crops cut and removed from the field than for the same crops pastured. This is simply because a considerable portion of the potash in the crop is returned to the soil through animal droppings. It should not be overlooked, however, that in many instances livestock, especially cattle, remain on the pasture for only a few hours a day and under such conditions the potash removed is almost as great as if the field had been cut for hay or silage.

Use higher potash rates for legumes

The rate of potash fertilization depends on the soil's ability to release available potash and on the proportion of the herbage composed of legumes. In the absence of legumes, the rate of potash fertilization can be reduced. Because of the "luxury" consumption of potash by legumes and grasses, this fertilizer plant food can normally be applied annually with better results than by making heavier applications less frequently.

7. Determining Deficiencies in Legumes

Legumes suffering from a deficiency of potash display very definite symptoms. On Ladino clover, white clover, red clover, and alfalfa, these symptoms show up as easily identified white spots on the leaves. A yellowing of the margin of the leaflet also occurs.

Observe the leaves

Many good farmers keep a close eye on the legumes in the mixture. As soon as any evidence of potash starvation shows on the leaves, they apply commercial potash. There is one precaution to keep in mind, however. Alfalfa plants infected with bacterial wilt may show

Comparing Fertilizer Costs per Nutrient Unit

COST PER UNIT OF NUTRIENT ELEMENT

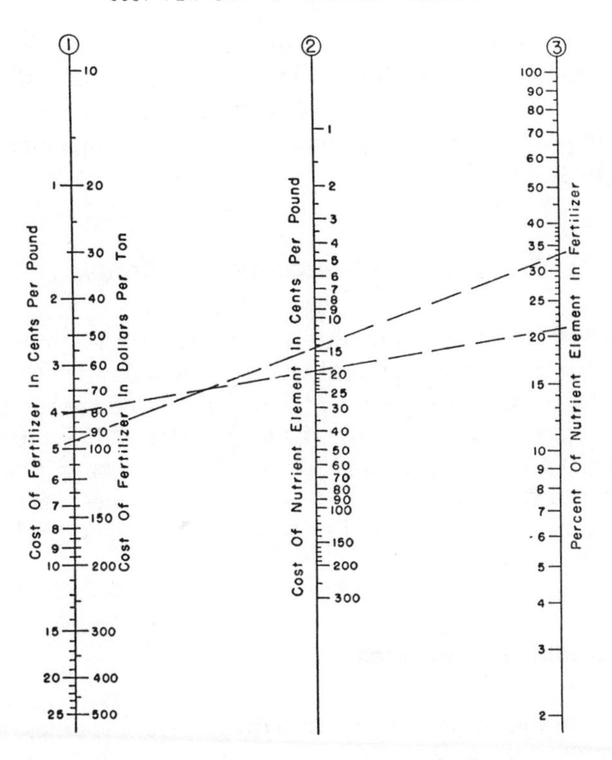

Courtesy, Oklahoma State University

Fig. 5.6—Use this "calculator" for figuring comparative plant-food costs. Line up a straight edge, one end at cost of fertilizer, the other end at percent of nutrient element in fertilizer, then take a reading from line 2 where the straight edge crosses it. This is the cost of plant food in cents per pound.

potash hunger signs even though the soil is adequately supplied with available potash. If plants infected with wilt show no signs of potash deficiency, the farmer can usually be sure that the level of available potash in the soil is sufficient for the crop's needs.

8. Using Rotational Grazing

Legumes need rotational grazing more than grasses. To keep legumes productive and competitive with grasses in a pasture, rotational grazing is desired. Rotational grazing cycles vary with species and climate. Alfalfa and red clover prefer short grazing periods and long rest periods. Ladino clover and birdsfoot trefoil tolerate longer grazing periods.

Rotational grazing is needed more in the humid regions than the arid or semi-arid regions. Humid region grazing of alfalfa should be 7 to 10 days of grazing with 30 days of rest. In semi-arid regions, alfalfa survives in pastures very well where 30 days of grazing and 30 days of rest are practiced.

9. Supplying Grasses with Nitrogen

Grasses require large amounts of nitrogen for high production. A considerable portion of the total amount needed can, of course, be obtained by including vigorous-growing, inoculated legumes in the mixture. In past years, authorities have emphasized that nitrogen fertilization of a good legume-grass sod was false economy. Even today, there is evidence that the use of nitrogen on legume-grass mixtures may do more harm than good unless the grazing is carefully controlled.

Avoid crowding out legumes

Nitrogen stimulates the growth of the grasses, and unless the grazing is carefully managed the legume is crowded or shaded out.

10. Boosting Yield of Grasses

On straight grass pasture, or on fields where the legumes have largely disappeared, commercial nitrogen, applied with an adequate amount of phosphate and potash, gives yields comparable to those obtained from legume-grass mixtures. On soils very low in available nitrogen, as much as 160 pounds of nitrogen (575 pounds of ammonium nitrate) applied in two or three applications will more than triple the carrying capacity of a pasture.

The use of straight grass pastures, heavily fertilized with nitrogen especially for very early spring, late fall, and winter grazing, is fairly common. Some farmers use straight grass pastures to eliminate the bloat hazard.

In recent years, the secondary and trace-element plant foods have received much attention. Today, most farmers prefer dolomitic limestone to high-calcium lime, since the former provides both calcium and magnesium. Many of our soils are deficient in magnesium.

Supply adequate sulfur

Sulfur deficiencies have been reported in some areas such as the West and Northwest. However, since large amounts of sulfur are returned to the soil through rain in the areas near industrial centers and in superphosphate and ammonium sulfate, sulfur deficiencies are not as widespread as those of nitrogen, phosphate, or potash. The student should keep in mind, however, that legumes use as much sulfur as phosphorus.

Use boron for legumes

Boron deficiency symptoms on legumes have been reported in all humid areas, especially on soils with less than 2 percent organic matter. Fertilizers carrying boron are recommended for legumes by many of our agricultural experiment stations.

11. Placing Fertilizer So It Can Be Reached by Plants

Most states recommend applying at least a portion of the fertilizer in a band below the seed. This puts the plant food close at hand where the small seedlings can pick it up. Thus they get off to a good start ahead of the weeds. When the fertilizer is broadcast throughout the seedbed it is just as available to the weeds as it is to the pasture seedlings.

Fig. 5.7—Every livestock product carries off valuable plant-food nutrients.

Band-place fertilizer

Most grain drills have band-seeding attachments, so it is possible to band-place fertilizer so there is a bit of soil between the fertilizer layer and the seed. This is important to keep the potash fertilizer from burning the seed and hurting the germination. Pennsylvania Agricultural Extension Service Leaflet 158 has this to say regarding band seeding of grasses and legumes:

> This system gives seeds shallow soil cover. Legume seeds should not be covered under a half-inch of soil.
>
> Fertilizer is placed so young seedlings reach it just after germination. If fertilizer is placed much over a half-inch from legume seeds the seedlings act as unfertilized plants.
>
> Legume and grass plants are fertilized; weeds are not.
>
> If band-seeding is performed properly the grass seed can be reduced by one-half and the legume seed by one-third under normal recommended rates.

Fig. 5.8—Many modern grain drills have band-seeding attachments.

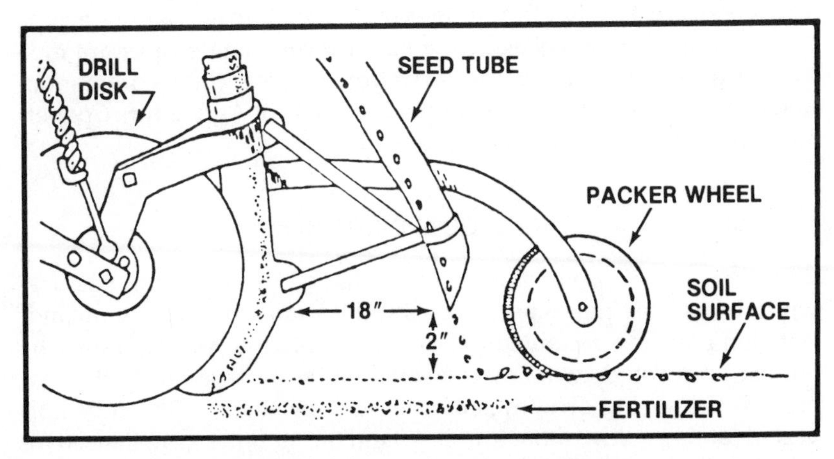

Fig. 5.9—Band-seeding is placing the seed on the soil surface directly above the band of fertilizer, which is placed 1-1½ inches deep in the soil.

12. Fertilizing for No-till Seeding

Many no-till seeders do not have fertilizer attachments. The seeders that do apply fertilizer place the fertilizer with the seed. Under most conditions, only phosphorus fertilizer should be placed with these seeders. Consequently, most of the fertilizer, especially potash and nitrogen, needs to be applied broadcast over the entire soil area. When seeding legumes or legume-grass mixtures into pastures with the no-till method, no nitrogen should be used on soils with 1½ percent organic matter or more. After newly seeded grasses are established and competitive with the old pasture grasses, nitrogen fertilization may begin.

13. Selecting a High-yielding Adapted Mixture

Chapter 4 gives information on the important grasses and legumes adapted to the different regions of the country. Farmers are confronted with selecting mixtures which will fit their climatic conditions; mixtures tailored to their livestock enterprises; and ones which they can manage with their farm labor load. There are unlimited combinations and each has its advantages and disadvantages.

It is suggested that the student write the extension agronomist at the state agricultural college for a list of recommended pasture mixtures. To show how mixtures vary within a state, the recommendations presented in University of Illinois College of Agriculture Cooperative Extension Circular 1186 follow.

ROTATION PASTURES

High animal production can be obtained from rotation pastures which consist of "hay-type" grasses and legumes grown on cropland. Rotation pastures, intended to last only two to four years, fit into the regular crop rotation, following small grain and preceding the row crop. If the spring growth is not needed for pasture, it can be harvested for hay or low-moisture silage and the summer growth used for supplementary pasture.

Grasses and legumes used in rotation pastures must be productive, palatable, nutritious, and easy to establish, but they need not be long-lived.

PERMANENT PASTURES

Much farm land should be left permanently in grass because of steep slopes or for other reasons. Pastures intended to occupy land eight or more years are classed as permanent pastures. Such pastures may be renovated with chemicals or by tilling, fertilizing, and reseeding from time to time, but they are seldom plowed out and planted to row or close-drilled crops. At least one of the grasses and one of the legumes in permanent pasture mixtures should be long-lived or a consistent reseeder.

Mixtures suitable to the Midwest are presented in Table 5.4.

Table 5.4

Forage Seed Mixture Recommendations for Illinois, in Pounds per Acre

For Hay Crops			
Central, Northern Illinois		**Central, Southern Illinois**	
Moderate to well drained soils			
Alfalfa	12	Alfalfa	8
		Orchardgrass	6
Alfalfa	8	Alfalfa	8
Bromegrass	6	Tall fescue	6
Alfalfa	8		
Bromegrass	4		
Timothy	2		
Alfalfa	8		
Timothy	4		
Poorly drained soils			
Alsike clover	5	Reed canarygrass	8
Timothy	4	Alsike clover	4
Reed canarygrass	8	Tall fescue	6
Alsike clover	3	Alsike clover	4
Birdsfoot trefoil	5	Redtop	4
Timothy	2	Alsike clover	4
Droughty soils			
Alfalfa	8	Alfalfa	8
Bromegrass	6	Orchardgrass	4
Alfalfa	8	Alfalfa	8
Tall fescue[1]	6	Tall fescue	6
		Alfalfa	8
		Bromegrass	6

(Continued)

Table 5.4 (Continued)

For Horse Pastures

Central, Northern Illinois		Southern Illinois	
Moderate to well drained soils			
Alfalfa	8	Alfalfa	8
Smooth bromegrass	6	Orchardgrass	3
Kentucky bluegrass	2	Kentucky bluegrass	5
Poorly drained soils			
Smooth bromegrass	6	Orchardgrass	6
Kentucky bluegrass	2	Kentucky bluegrass	5
Timothy	2	Ladino clover	½
Ladino clover	½		

Central Illinois

Moderate to well drained soils	
Alfalfa	8
Orchardgrass	3
Kentucky bluegrass	2
Poorly drained soils	
Orchardgrass	6
Kentucky bluegrass	2
Ladino clover	½

For Pasture Renovation

Central, Northern Illinois		Southern Illinois	
Moderate to well drained soils			
Alfalfa	8	Alfalfa	8
Red clover	4	Red clover	4
Poorly drained soils			
Birdsfoot trefoil	4	Alsike clover	2
Red clover	4	Ladino clover	½
		Red clover	4

For Rotation and Permanent Pastures

Central, Northern Illinois		Southern Illinois	
Moderate to well drained soils			
Alfalfa	8	Alfalfa	8
Bromegrass	5	Orchardgrass	4
Timothy	2	Alfalfa	8
Alfalfa	8	Tall fescue	6
Orchardgrass[1]	4	Tall fescue	8
Alfalfa	8	Ladino clover	½
Orchardgrass[1]	4		
Timothy	2		

(Continued)

Table 5.4 (Continued)

Mixture	Rate
Orchardgrass[1] Ladino clover	6 ½
Red clover Ladino clover Orchardgrass[1]	8 ½ 4
Red clover Ladino clover Tall fescue	8 ½ 6-8
Birdsfoot trefoil Timothy	5 2
Bromegrass Ladino clover	8 ½
Tall fescue	10
Orchardgrass[1]	8
Alfalfa Bromegrass Timothy	8 6 2
Orchardgrass Ladino clover	6 ½
Tall fescue	10
Orchardgrass	8
Red clover Ladino clover Orchardgrass	8 ½ 4
Red clover Ladino clover Tall fescue	8 ½ 6-8
Poorly drained soils	
Alsike clover Ladino clover Timothy	3 ¼ 4
Birdsfoot trefoil Timothy	5 2
Reed canarygrass Alsike clover Ladino clover	8 3 ¼-½
Alsike clover Tall fescue Ladino clover	2 8 ½
Alsike clover Tall fescue Ladino clover	2 8 ½
Reed canarygrass Alsike clover Ladino clover	8 3 ½
Droughty soils	
Alfalfa Bromegrass	8 5
Alfalfa Orchardgrass[1]	8 4
Alfalfa Tall fescue	8 6
Red clover Ladino clover Orchardgrass	8 ½ 4
Red clover Ladino clover Tall fescue	8 ½ 6-8
Alfalfa Orchardgrass	8 4
Alfalfa Tall fescue	8 6
Red clover Ladino clover Orchardgrass[1]	8 ½ 4
Red clover Ladino clover Tall fescue	8 ½ 6-8

For Hog Pastures

All soil types, anywhere in Illinois

Mixture	Rate
Alfalfa Ladino clover	8 2

[1]Central Illinois only.

ANNUAL PASTURE CROPS

Annual pasture crops occupy the land for only one year, although such an annual mixture as wheat-lespedeza may be grown repeatedly on the same field. Examples of annuals for pasture and their seeding rates are:

1. Rye 2 bushels per acre.
2. Rye 1½ bushels and hairy vetch 15 or crimson clover 10 pounds per acre.
3. Rye 1 bushel, ryegrass 10 pounds, and hairy vetch 15 or crimson clover 10 pounds per acre.
4. Winter wheat 1½ bushels per acre.
5. Winter wheat 1 bushel and crimson clover 10 pounds per acre.
6. Winter barley 2 bushels per acre.
7. Winter barley 1½ bushels and crimson clover 10 pounds per acre.
8. Winter oats 3 bushels per acre.
9. Winter oats 2 bushels and crimson clover 10 pounds per acre.
10. Annual lespedeza (in grain crop) 25 pounds per acre.
11. Sudangrass 25 pounds per acre.

14. Planting in Spring and Late Summer

Early seeding desirable

In the humid areas of the United States, more difficulty is encountered from weeds and old, unkilled vegetation in spring seedings than in late summer seedings. On the other hand moisture in the seedbed is usually greater in the spring. These conditions assure faster germination of the small seeds. Young plants started in the spring become well established before the cold winter months set in and are not as likely to winter-kill as late summer seedings. Early spring seedings have fewer weed problems than late spring seedings.

Late summer seedings

Late summer seedings are equal or superior to spring seedings in the southern half of the Corn Belt. Seedbeds can be well-prepared during a dry period shortly before seeding. The rainfall distribution in the southern Corn Belt favors late August to early September seedings. Summers usually have low amounts of rainfall, and the probability of rainfall increases as fall approaches. Seedings made on a well-prepared, firm seedbed in late summer can expect to have increasing frequencies of rainfall for seed germination and the support of seedling growth.

Most troublesome weeds to pasture seedings are annuals, and many are warm-season annuals. Late summer is when these weed species *stop* growing, rather than *begin* growing, as in the spring. Thus, late summer seeded pastures usually have fewer weed problems than spring seeded pastures.

The time of seeding in late summer is determined by geographical location. Late summer seedings should be made by August 25 in southern Iowa and by September 25 in southern Illinois, for example.

15. Using a Companion Crop

A companion crop of oats, wheat, barley, or rye can help hold down weeds, lessen soil erosion losses, and, of course, provide a cash crop. On the other hand, a companion crop offers severe competition for existing soil moisture, plant nutrients, and sunlight. In deciding whether to use a companion crop, the farmer should consider:

1. The need for a grain crop or straw for bedding.
2. Whether weeds are likely to be a serious problem.
3. Whether the erosion hazard may be present during the establishment period.
4. Whether the forage crop can tolerate shading in the seedling stage. For example, alfalfa does not tolerate shading in the seedling stage as much as red clover.

Reduce seeding rate

When spring grain is used as a companion crop, the seeding rate of the grain should be about one-half the usual recommended rate for grain production. When the spring grain is harvested as a forage crop, the normal seeding rate for grain production is used.

When early establishment and rapid growth of forage are of pri-

Fig. 5.10—Spring oats is a good companion crop for pasture seedings.

mary consideration, and erosion and weeds are not considered serious, seeding without a companion crop is recommended.

At the U.S. Department of Agriculture Regional Pasture Laboratory in Pennsylvania, alfalfa alone and in a mixture of red clover–Ladino, without a companion crop of oats, yielded more than where the companion crop was used. This was true even though the oats crop was cut for hay and was included in the total yield. On the other hand, total yield of a later seeding made May 15 was increased by the companion crop. Most of the total yield, however, consisted of oats. These experiments emphasize the importance of early seeding. When a late summer crop is made, a companion crop is usually not recommended.

Some farmers use a light planting of a spring small grain like oats when making late summer seedings. The small grains germinate rapidly; their seedlings grow rapidly and hold the soil in place until the pasture seeding is established. When cold weather comes, the oats are killed, but the dead plants still afford some winter protection to the young grasses and legumes, and some soil cover that reduces soil erosion.

16. Obtaining Quality Seed

The farmer should always be certain to plant only high-quality seed of adapted grasses and legumes. The cheapest seed is often the most expensive in the long run. Seed firms, farm advisors, and reputable dealers are good sources. Check the label. Germination tests are required on all seed sold through wholesale and retail outlets. Purity and germination are two very important features of high-quality seed.

17. Inoculating Seed

The importance of inoculating the legumes in the mixture should not be underemphasized. Unless the legumes are inoculated they cannot fix atmospheric nitrogen from the air but will compete with the grasses for the nitrogen being released by the soil. Under such conditions, yields will be reduced and forage quality will be poor.

Fig. 5.11—A good balance between grass and clover.

Follow these directions for inoculating legume seed

1. Controlled moisture.
   Place seed in mixing or seeding container.
   Apply sufficient volume of inoculant and mix thoroughly.
   Seed at once unless stored in a cool place.
2. Added moisture.
   Place seed in mixing container.
   Sprinkle with water, blending seed so that all are moistened, not wet.
   Apply sufficient inoculant and mix thoroughly.
3. Slurry and seed mixers.
   Make certain all seeds are coated by thoroughly mixing.

18. Preparing the Seedbed

Regardless of the mixture chosen, it should be seeded on a firm seedbed. A firm seedbed is important in the germination and early es-

Courtesy, USDA

Fig. 5.12—Legumes are heavy users of phosphate and potash. Phosphate fertilizer made this difference.

tablishment of a new seeding, the reason being that the compaction of the seedbed increases the capillary porosity of the soil, thereby increasing the available water. Moreover, soil moisture will move by capillary action through a firm seedbed and will keep the surface soil moist for a considerable period. The most common tool used for preparing a firm seedbed is the cultipacker or the cultipacker-seeder.

19. Planting Seed

Do not place deep

The Ohio Agricultural Experiment Station has found that about ¼ inch to ½ inch deep on non-mulched land is usually best. Information on seed emergence when planted at different depths is presented in Table 5.5.

Table 5.5

Relationship of Planting Depth to Seed Emergence[1,2]

	Percentage of Emergence When Sown at a Depth of:					
	¼″	½″	1″	¼″	½″	1″
Crop	Not Mulched			Mulched—1 Ton Straw		
Yellow sweet clover	44	43	30	78	78	73
White sweet clover	34	30	15	60	62	42
Alfalfa	75	63	48	85	82	73
Red clover	44	39	25	85	86	70
Alsike clover	21	22	7	91	91	46
Crimson clover	83	78	49	95	93	75
Korean lespedeza	88	82	33	96	92	66
Average, all legumes	56	51	30	84	83	63
Orchardgrass	58	59	40	95	86	69
Timothy	46	27	3	70	68	23
Sudangrass	64	78	75	80	74	84

[1]*Source: Handbook of Ohio Experiments in Agronomy.*

[2]Seed sown May 18.

20. Managing the First Year

Remove companion crops early

Equally as important as getting a good stand of grasses and legumes is managing the stand to favor satisfactory growth and persistence through the first winter. When a small grain companion crop is used, best results are obtained by removing it about the time it comes in head. If a grain crop is harvested it should be harvested just as soon as possible and the straw removed from the field.

The plants must make a vigorous growth and become well established before winter sets in if they are to survive. This means that under unfavorable growing conditions spring seedings should not be grazed the first year. On the other hand, if moisture conditions have been good, a spring seeding without a companion crop may furnish light grazing in July, August, and early September. If a companion crop is used the growth will be retarded and possible grazing usually will be delayed about a month.

Don't overgraze the new seeding

Care must be taken not to graze the new seeding too closely or too long. Remember that the tops feed the roots and unless the top growth is permitted to grow vigorously the root system will be weak and plants will winter-kill.

Mow weeds

On fields previously in row crops, weeds may become serious, particularly where a companion crop was not used. In such cases the weed and forage growth should be cut with a mower and removed about the time the alfalfa or red clover in the mixture begins to bloom. Under severe weed competition, even earlier clipping may be desirable. In cutting the weed-forage mixture it is usually best to set the cutter bar of the mower low. This retards the recovery of the weeds more than that of the new growth of grasses and most legumes (sweet clover being an exception). The new growth buds of most grasses and legumes are near ground level, whereas on most annual weeds the buds are carried above ground on the main stem.

Fig. 5.13—Weeds can be controlled by herbicides.

In most instances the weed-forage mixture can be used to make nutritious silage.

21. Increasing Legumes in Grass Pastures

Increasing legumes in grass pastures in order to improve the protein content of the forage and extend the grazing season may be a major problem without a renovation practice. Proper fertilization can be used to increase the legume yield more than the grass yield. Potassium fertilization will benefit legumes more than grasses. If the soils are acid, adding limestone will encourage legumes more than grasses. Phosphorus fertilization on deficient soils will also aid legumes more than grasses. Proper fertilization and rotational grazing can increase the productivity of the legumes in a pasture and increase the longevity of those legumes in the pasture.

21. Increasing Legumes in Grass Pastures

CHAPTER 6

Understanding Forage Plant Nutrient Requirements

More farmers are learning that applications of limestone and fertilizers are essential to the establishment of forage crops initially and that top-dressing is necessary for high yields and long-lived stands. The mere spreading of fertilizer is not enough since the time of application and the ratio of nutrients applied are important considerations in pasture management. Grasses often go dormant in summer because of a shortage of nitrogen, and legumes in a grass-legume mixture may lose out because the grasses feed more efficiently on potassium. Since the application of nitrogen alone may accentuate this relationship, the application of potassium concurrently and perhaps phosphorus initially must receive consideration when nitrogen applications are appropriate. A balanced fertilizer application program should increase yields, improve forage quality, and encourage persistence of desirable species.

Then, too, timely fertilizer application should stimulate earlier regrowth of forage plants. The proper balance of nitrogen and potassium, particularly in top-dressing, should lead to more efficient protein synthesis and less accumulation of incompletely metabolized forms of nitrogen in forage plants, apparently a factor affecting feed intake.

The advisability of spreading fertilizers to correct nutrient deficiencies may have considerable significance in forage production in addition to resulting in greater yields of dry matter. It is hoped that the reader will have a better comprehension of forage plant nutrient requirements and a better understanding of the whys of fertilizers and limestone applications from the brief presentations that follow in this chapter.

Activities Which Involve Approved Practices

1. Recognizing plant nutrient deficiency symptoms.
2. Recognizing deficiency symptoms in grass plants.
3. Recognizing deficiency symptoms in legume plants.
4. Original sources, current sources, and functions of plant nutrients.

1. Recognizing Plant Nutrient Deficiency Symptoms

Plants show hunger signs which can be detected by careful observation and study. The deficiency symptoms are definite, but the proper treatment may require considerable experience. However, tissue testing, soil testing, and plant nutrient deficiency symptoms are sufficiently well correlated to enable one to diagnose the cause of low yields of forage.

If lime and fertilizers have been applied according to soil tests, deficiency symptoms in forage plants should not show up, but a part of pasture management should be the ability to recognize *nitrogen* deficiency, *phosphorus* deficiency, *potassium* deficiency, and perhaps other plant nutrient deficiencies, if they should occur.

In general, plants hungering for *nitrogen* display a sickly yellowish-green color, a slow and dwarfed growth, and a "firing" (scorching) of leaves, starting at the bottom of the plant and proceeding upward, with the "firing" of the individual leaf beginning at the tip and proceeding down the center or midrib.

Phosphorus deficiency generally results in a purplish coloration of leaves, stems, and branches; slow growth and maturity; lack of tillering in grasses; and low yields of grain and seed.

In many plants *potassium* deficiency causes a "scorching" of margins of lower leaves, starting at the tip and proceeding down the edges of the individual leaf, leaving the midrib reasonably green. Premature loss of leaves may occur, and the plants may fall over due to lack of stem stiffness and inadequate root development.

2. Recognizing Deficiency Symptoms in Grass Plants

1. Nitrogen—Grass plants needing more nitrogen generally have purplish-green, spindly stems and yellowish leaves, the tissues of which die from the tip downward along the midrib.

2. Phosphorus—Grass plants hungering for phosphorus appear dark green and otherwise healthy but exhibit slow growth, poor root development, and poor tillering.

3. Potassium—The leaves of grass plants short on potassium display a "scorching" of the leaves, starting at the tips and proceeding down the edges in the early stages, after which the leaves turn yellow, then brown, and die. A shortage of potassium results in weak stem development.

4. Magnesium—In severe magnesium deficiency, a yellow to white streaking between the parallel veins of the grass leaves develops and extends the full length of the blade. These streaked tissues may dry and eventually die.

5. Sulfur—Sulfur deficiencies are more commonly found in legumes than in grasses. The plants are yellowish-green in color and grow slowly. In many respects, a sulfur deficiency resembles that of nitrogen deficiency with the exception of the midrib V-pattern.

3. Recognizing Deficiency Symptoms in Legume Plants

1. Nitrogen—A deficiency of nitrogen results in slower growth, less branching of the plant, and smaller plants at maturity. A pale green to chlorotic condition tends to spread over the whole leaf surface with affected foliage showing up simultaneously at the base tips of branches and tip of the main stem. If observed early enough, an application of nitrogen fertilizer will restore the normal green color.

2. Phosphorus—A deficiency of phosphorus in legume plants results in slow growth, with the plants remaining poorly developed and a delayed flowering and seed production. Seedling legumes often have purple coloration of the leaves.

3. Potassium—In alfalfa, sweet clover, and the clovers generally, a shortage of potassium results in the appearance of yellowish dots at the tips and near the margins of leaves on all parts of the plant. (These symptoms should not be confused with leaf hopper injury.) The yellowish mottling continues around the edges of the leaflets, with the dead tissue falling out, giving the leaflets a ragged edge appearance.

4. Magnesium—A shortage of magnesium in legume plants

shows up in yellowing to chlorotic patches between the main veins of the leaflets, with the areas along the veins remaining green.

5. Manganese—If manganese deficiency symptoms occur in legume plants, they will be similar to magnesium deficiency symptoms, except that they are more apt to appear on the upper, younger leaflets.

6. Boron—A deficiency of boron results in a "whitening" of the terminal buds of legume plants. In severe shortages, the internodes do not elongate normally and produce a condition known as "crazy top," where leaf whorls all seem to originate from the same position around the main stem.

7. Sulfur—A sulfur-deficient plant is yellowish-green, as contrasted with a dark green plant that is well supplied with both nitrogen and sulfur. Legumes need more sulfur than do grasses. Plants low in sulfur are usually low in protein.

Other plant nutrient deficiency symptoms have been observed, but they are not definite enough in coloration to be reported here.

Table 6.1

Kinds of Plant Organisms Found in Soils

I. Bacteria—500 lbs./acre
    A. Heterotrophic
        1. Nitrogen fixers
            a. Symbiotic
            b. Non-symbiotic
                (1) Aerobic
                (2) Anaerobic
        2. Those requiring fixed nitrogen
    B. Autotrophic
        1. Nitrite formers
        2. Nitrate formers
        3. Sulfur oxidizers
        4. Iron oxidizers
        5. Those that act on hydrogen and other hydrogen compounds

II. Fungi—1,000-1,500 lbs./acre
    A. Yeast and yeastlike fungi
    B. Molds
    C. Mushrooms

III. Actinomyces—700 lbs./acre

IV. Algae
    A. Blue-green
    B. Grass green
    C. Diatoms

4. Original Sources, Current Sources, and Functions of Plant Nutrients

Nitrogen

1. Original sources—fixation of nitrogen from the air by symbiotic and free-living bacteria, as well as electrical fixation of nitrogen from the atmosphere (79 percent nitrogen). This has resulted in plant growth that through eons of time has developed the humus and organic matter of soils containing up to 5 percent nitrogen (higher for mucks and peats).

2. Current sources—soil humus by nitrification, anhydrous ammonia, ammonia solutions (and other nitrogen-containing chemical compounds).

3. Functions—needed throughout life of plant.

Is an essential constituent of amino acids and proteins of protoplasm.

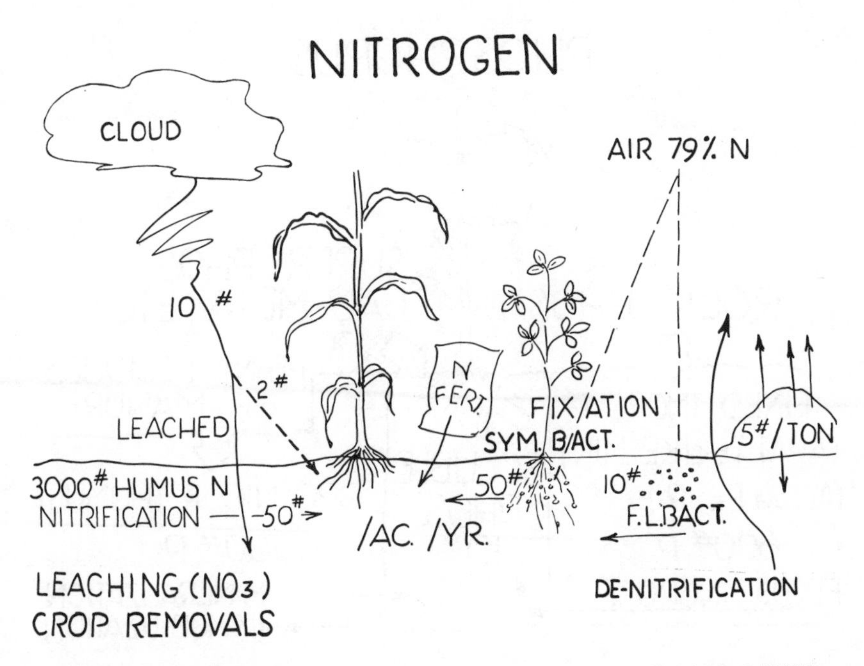

Drawing by C. McVickar

Fig. 6.1—Schematized nitrogen cycle.

Feeds microorganisms during decomposition of low-nitrogen organic materials.

Forms an integral part of the chlorophyll molecule.

Is a part of every cell and promotes rapid growth and quality of leaf crops.

Phosphorus

1. Original sources—apatite, poly-calcium mineral, with varying amounts of halogens.
2. Current sources—metaphosphate, nitric phosphates, treble and single superphosphate, rock phosphate (mono-, di-, and tri-calcium phosphates), and ammonium phosphates.
3. Functions—needed early in plant growth for root development and later when grain is filling (for cell division).

Stimulates flowering.

Aids transforming starches to sugars.

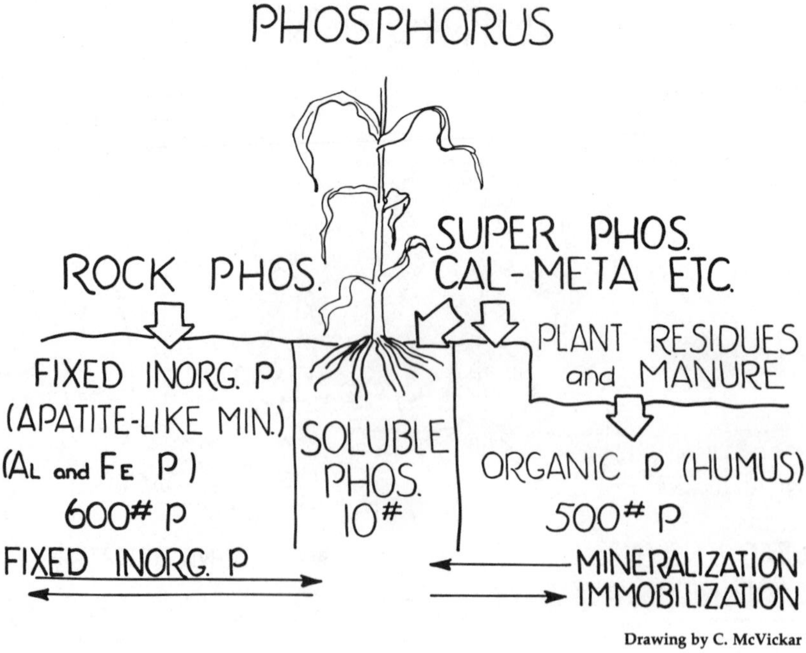

Drawing by C. McVickar

Fig. 6.2—Schematized phosphorus cycle.

Is essential for seed formation (hastens maturity).

Certain high-energy phosphate bonds are active in photosynthetic and respiratory processes.

Gives winter hardiness to fall seeded grains and forage crops.

Potassium

1. Original sources—micas and potassium feldspars.
2. Current sources—muriate of potash and some sulfate and nitrate of potassium fertilizers.
3. Functions—needed throughout life of plant because it is involved in translocation of food and structural materials.

Is a catalyst in photosynthesis and necessary for development of chlorophyll.

Imparts increased vigor and disease resistance to plants.

Promotes stem and leaf growth.

Is involved in metabolism of nitrogen in plants.

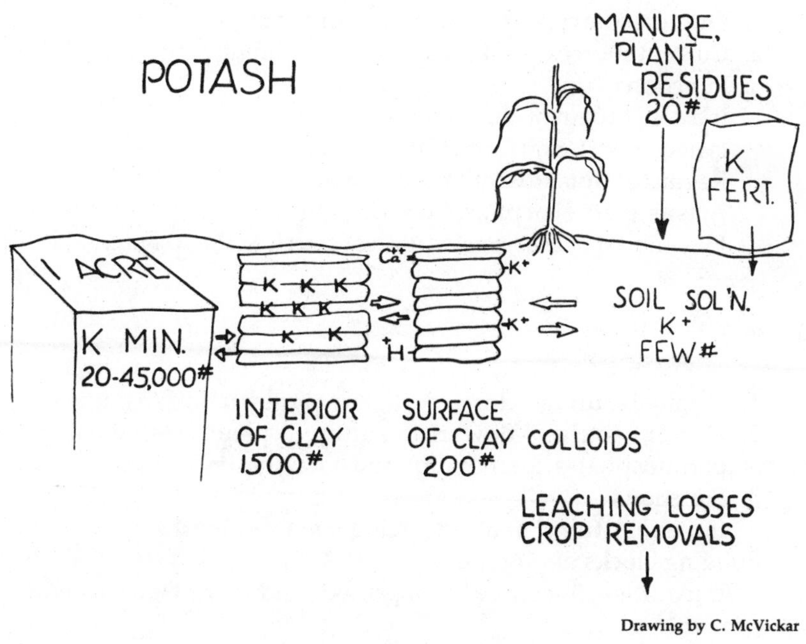

Drawing by C. McVickar

Fig. 6.3—Schematized potassium cycle.

Is essential to formation and transfer of starches, sugars, and oils.
Imparts winter hardiness to legumes and other crops.

Calcium

1. Original source—calcium carbonates.
2. Current sources—calcium carbonates, oxides, and hydroxides (liming materials).
3. Functions—
Regulates permeability of root hairs to other ions (intake of other nutrients).
Promotes plant vigor and stiffness of stem.
Increases calcium content of feed crops.
Promotes growth of legumes.

Magnesium

1. Original sources—dolomite, talc, serpentine.
2. Current sources—dolomite limestone (liming material).
3. Functions—
Is a constituent of chlorophyll.
Is necessary for oil formation in plants.
Regulates uptake of other nutrients.
Acts as a carrier of phosphorus in plants.
Plays a part in the translocation of starch within plants.

Sulfur

1. Original sources—flowers of sulfur, pyrites, and gypsum.
2. Current source—flowers of sulfur, gypsum (residual sulfur from superphosphates, ammonium, and potassium sulfates).
3. Functions—
Is a constituent of amino acids (cystine, methionine)—the building blocks of protein.
Encourages increased root growth and more vigorous plant growth.
Is essential for nodule formation on legume roots.

Sulfide linkage is involved in conversion of radiant energy of sun to chemical energy, which is stored as food in plants.

Boron

1. Original source—tourmaline.
2. Current source—borax and slow-release compounds.
3. Functions—
   Plays a role in protein synthesis.
   Is related to both calcium and potassium metabolism in plants.
   Is necessary for nodule formation in legumes.

Iron

1. Original sources—iron oxides and hydroxides.
2. Current source—ferrous sulfate and ferric sulfate.
3. Functions—
   Is a catalyst in chlorophyll formation.
   Acts in conjunction with certain respiratory enzyme systems of plants.

Manganese

1. Original sources—magnesian rocks and minerals.
2. Current source—manganese sulfates and manganese oxide.
3. Functions—
   Is necessary for the functioning of certain oxidative enzyme systems.
   Is an oxidant of iron (makes iron more available to plants).

Zinc

1. Original source—sphalerite.
2. Current source—sulfate of zinc, zinc oxide, zinc nitrate, and zinc ammonium nitrate.
3. Function—
   Is associated with the carbonic anhydrase and carboxylase enzyme systems of plants.

Table 6.2

Plant Food Elements Removed from Soil per Crop Unit

	Nitrogen	Phosphorus	Potassium	Calcium	Magnesium	Sulfur
	(pounds)					
Alfalfa hay, 8 tons	360	36	300	300	60	38
Bluegrass hay, 2 tons	54	8	70	16	16	12
Clover hay, 4 tons	165	15	110	116	18	24
Soybean hay, 4 tons	200	24	70	62	10	42
Timothy hay, 1.5 tons	30	4	35	14	5	3
Lespedeza hay, 1 ton	43	5	16	20	6	—
Good clover-grass pasture that supports cow and calf per acre	190	25	200	100	30	18
Sweet clover						
Annual, 2 tons	100	10	56	44	12	16
Biennial, mature, 4 tons	160	14	150	70	32	24
Biennial, second spring, 2 tons	150	10	70	60	18	18
Barley (grain), 60 bu.	54	10	20	2	2	4
(straw), 1.6 tons	18	2	34	10	2	8
Oats (grain), 100 bu.	64	10	16	4	4	6
(straw), 2 tons	24	4	50	14	4	8
Rye (grain), 60 bu.	64	10	16	2	2	4
(straw), 3 tons	28	8	40	20	4	60
Wheat (grain), 50 bu.	56	20	12	2	4	4
(straw), 2 tons	20	4	24	8	4	—
Corn (ear), 120 bu.	120	20	36	2	6	12
(stover), 4 tons	72	18	80	30	6	12
1,000 lbs. of milk taken from farm	6	2	2	2	—	—
1,000 lbs. of beef taken from farm	27	8	2	13	—	—

Molybdenum

1. Original source—molybdenite.
2. Current source—sodium or ammonium molybdate (trace mineral of fertilizers).
3. Functions—

   Is necessary for functioning of nodule bacteria.

   Assists the functioning of nitrate-reducing enzymes.

CHAPTER 7

Keeping Pastures Productive and Healthy

Once the livestock farmer has established a fine productive pasture, his or her interest turns to keeping the pasture producing efficiently for as many years as possible. It costs money to renovate because cash outlays are necessary to buy seed and fertilizer. However, investments for both items are worthwhile. Moreover, lengthening the productive life of a pasture makes production of livestock more efficient.

Activities Which Involve Approved Practices

1. Supplying plant food to pastures.
2. Maintaining the grass-legume ratio.
3. Preventing leaching of nitrogen.
4. Liming pastures.
5. Checking the lime level.
6. Controlling grazing.
7. Preventing severe grazing.
8. Clipping pastures.
9. Handling poisonous plants in pastures.

Adequate fertilizer and lime, controlled grazing, and regular clipping to control weeds and scatter dung piles are major practices which largely determine whether the pasture will have a long, profitable life or whether it will dwindle away before it has yielded a satisfactory return.

1. Supplying Plant Food to Pastures

Few appreciate the large amounts of plant food nutrients used by vigorous-growing pastures. We are told that the herbage consumed by a mature cow, during a six-month grazing period, carries approximately 125 pounds of nitrogen, 25 pounds of phosphoric acid, and 100 pounds of potash.

Manure contains nutrients

The manure droppings, of course, will return a portion of the plant food, but even so, there is a constant drain from the soil. For example, 8,000 pounds of milk contains the equivalent phosphorus found in 100 pounds of 20 percent superphosphate. Every livestock product that leaves the farm carries off plant food nutrients. In the skeleton of a 1,000-pound steer, there is also the amount of phosphorus equal to that found in a 100-pound bag of 20 percent superphosphate.

Balance soil deficiencies

It should be pointed out, however, that during the soil weathering process certain quantities of the essential plant foods are released from the more insoluble minerals. So, in practice, the farmer tries to balance any soil deficiency by using fertilizer and lime to make up the difference between what is needed by the crop and what is given up by the soil.

In a later chapter we will discuss the effects of fertilization on the nutritive value of pasture herbage.

Apply commercial fertilizer

The farmer is interested in applying commercial fertilizer as long as the additional livestock product obtained pays well for the money spent for this investment. It is easier to make this statement than it is to prescribe a method for determining how much fertilizer will pay off. Theoretically, the amount that can be profitably used is an amount

Courtesy, New Holland Division of Sperry Rand

Fig. 7.1—Fertilizers are a profitable investment for pastures.

equal to the difference between the quantity released by the soil and that needed by the pasture for near-top production.

Soil tests have already been discussed in previous chapters. Their use every three or four years is recommended for determining the nutritive status of soils growing established pastures.

Experiment on your farm

Many progressive farmers also experiment on their own farms to determine the kinds and amounts of fertilizer that are best for their pasture programs.

2. Maintaining the Grass-Legume Ratio

One of the reasons why it is so difficult to make general pasture fertilizer recommendations is that any shift in the grass-legume ratio of the pasture mixture greatly influences the plant-food nutrient up-

Fig. 7.2—Legumes increase the protein content of the herbage.

take from the soil. Legumes, for example, properly inoculated, can provide their own nitrogen requirements through the symbiotic association with the nodule bacteria on the roots. In fact, legumes can also release nitrogen to the companion grass crop. Table 7.1 gives the average amount of nitrogen fixed by some of our more important legumes.

The following approved practices in pasture management favor the legumes:

1. Grazing close (2 inches or less) to insure full use of forage by livestock.
2. Rotating grazing with enough animals per acre to graze down in a week to 10 days.
3. Clipping to control weeds and patches of over-ripe grass.
4. Avoiding undergrazing which favors the grasses and wastes forage.
5. Removing excess forage by hay making if pasture gets ahead of the animals. This keeps forage young and tender and insures ready feed for stress periods in winter or summer.

Table 7.1

How Much Nitrogen Do Legumes Fix?[1]

Legume	Average Pounds of Nitrogen per Acre
Alfalfa	194
Sweet clover	119
Red clover	114
Ladino clover	179
White clover	103
Alsike clover	119
Crimson clover	94
Lespedeza (annual)	85
Soybeans	58
Winterpeas	50
Vetch	80
Lupines	151
Kudzu	107

[1]*Source:* U.S. Department of Agriculture Farmers' Bulletin 2003.

Typical benefits reaped by keeping the legumes in the pasture mixture are illustrated by Indiana data. Bluegrass with 120 pounds of nitrogen per acre produced 262 pounds of beef, while bluegrass plus birdsfoot trefoil gave 297 pounds of beef and had a greater carrying capacity.

3. Preventing Leaching of Nitrogen

The student should keep in mind, however, the fact that nitrogen is a moveable plant food nutrient and can be lost to depths below the root zone by leaching. Phosphate moves very little from where it is placed, and this explains why many authorities suggest putting down enough phosphate at seeding time to last the life of the pasture mixture. Potash is intermediate between nitrogen and phosphate, although it doesn't leach nearly as readily as nitrogen. Potash is liked by legumes. When surplus quantities are put on in any one application, legumes take up far greater quantities than are needed for normal growth. Since the livestock require very small amounts of potassium, it is false economy to have luxury consumption by the herbage.

Fertilize in fall or spring

Most of our agricultural colleges recommend either fall or spring fertilization. Authorities who recommend fall fertilization point out the following advantages:

1. Fertilizer applied in the late summer or early fall builds food storage within plants, making them more winter hardy.
2. The land is in better condition for making the fertilizer application in the fall than it is in the spring. It usually is not wet and soggy.
3. Farmers usually have more time during the fall months than during the rush spring months.

On the other hand, those who advocate spring application point out that there may be considerable losses of plant food, especially nitrogen and some potash, from fall applications as a result of leaching during the winter months.

This difference of opinion among some of the authorities has led many of our outstanding grassland scientists to suggest that the important thing is to apply the fertilizer and not be especially concerned about the time of application.

Place fertilizer shallow

There is a difference of opinion as to the best method of fertilizing. There are those who reason that the fertilizer should be applied by some machine which places it below the soil surface and in the root zone.

In general, experimental data do not justify this recommendation. Experimental work at the Ohio station showed no advantage in the placement of fertilizer more than 2 inches deep. In a wet year, surface application was preferable.

Likewise, Indiana data showed surface application was superior to placing fertilizer in bands 4 to 8 inches apart at a depth of 2 inches in a bluegrass sod. Michigan experiments showed no difference for surface or below-turf placement.

Employ custom application

With the advent of the truck-broadcast spreader, it is much easier for the farmer to have the fertilizer custom spread on the established pastures than to apply it himself or herself with a machine that puts it down in the soil.

The National Joint Committee on Fertilizer Application, an organization no longer in existence but whose sponsors were The American Society of Agronomy, The American Society of Horticultural Science, The American Society of Agricultural Engineers, The Farm Equipment Institute, The National Canners' Association, and the National Plant Food Institute, made the following recommendations in a publication, "Methods of Applying Fertilizer."

> *The fertilizer should be applied with a grain drill or as a broadcast top dressing in either fall or spring. Pastures in the South that are used for winter grazing should be fertilized in late summer or early fall.*
>
> It is generally agreed that when a desirable sod is already established or where it is undesirable to appreciably disturb the soil, top dressings are satisfactory. Some prefer to apply the fertilizer in bands, slightly grooving it into the soil, as with a disc drill. Placement at a depth greater than 2 inches appears to be less satisfactory than either surface or shallow treatment. Applications of potash, on soils deficient in this element, should be made annually or every two years for best results.
>
> Broadcast or drilled applications of nitrogen on the surface of the soil in either the fall or early spring have generally given satisfactory results. There appears to be little need for incorporating the nitrogen into the soil, except where such special forms as liquid ammonia are used.

In general the quantities of fertilizer needed are far greater than those being used currently by farmers.

The average rate of fertilization of grassland (both pastures and hayland) is given in Table 7.2.

Apply by air

One very rapid and efficient way to apply plant nutrients is by use of the airplane. Air application is especially useful on rangelands; on rough, rocky soils; or under wet, soggy conditions where field equipment cannot go. The timing of fertilizer application can be a

Table 7.2

Our Grasslands Are Poorly Fertilized[1]

Region	Tame Hay and Cropland Pasture	Average Quantity of Fertilizer Used per Acre	Total Quantity of Fertilizer Applied	Average Quantity of $N–P_2O_5–K_2O$ Used per Acre	Total Quantity of $N–P_2O_5–K_2O$ Applied	Percentage of Acres Fertilized	Average Quantity of Fertilizer Used per Acre Where Fertilized
	(1,000 acres)	*(pounds)*	*(tons)*	*(pounds)*	*(tons)*	*(%)*	*(pounds)*
Northeast	9,254	69.9	323,227	22.8	105,291	19.1	366
Lake States	13,415	26.6	178,118	10.5	70,665	11.0	241
Corn Belt	23,004	32.9	378,270	9.8	112,287	11.0	299
Northern Plains	12,195	12.4	75,666	5.6	34,215	7.7	161
Appalachian	14,472	75.6	546,927	23.7	171,603	17.6	429
Southeast	4,845	204.5	495,411	43.7	105,898	38.0	538
Delta	6,441	60.2	193,975	17.7	57,036	19.0	317
Southern Plains	11,990	39.4	236,375	13.2	79,329	16.5	239
Mountain	10,561	24.5	129,306	9.4	49,736	13.4	183
Pacific	7,227	75.0	270,956	24.6	88,789	29.6	253
48 States	113,404	49.9	2,828,231	15.4	874,848	15.7	318

[1]*Source:* John Gale, USDA.

critical factor in its effectiveness, and the airplane can do the job under most weather conditions. Cost of application is competitive, especially on large acreages.

4. Liming Pastures

Recommendations for applying lime at the time of pasture establishments have already been discussed in Chapter 5. However, for pastures to remain productive it may become necessary to make additional applications of lime.

Work in lime

The initial lime application, made prior to seedbed preparation, should be thoroughly worked into the root zone. If this has been done, surface applications will usually suffice for proper maintenance.

Add enough lime to last for several years

Under most soil conditions, it is feasible to add enough lime at establishment time to take care of the soils and crop needs for usually four to six years. The merits of putting on heavier applications less frequently are demonstrated by research at the Ohio Agricultural Experiment Station.

The following is a direct quotation from the *Handbook of Ohio Experiments in Agronomy:*

> A 26-year test conducted on soils of rather low productivity at the Northeastern Experimental Farm in the southern part of Cuyahoga County (the Cleveland area) shows that there is no reason for applying small quantities of liming material at frequent intervals. When the total amount used over a period of years was the same, it was somewhat better to apply a large initial amount and then skip a rotation or two rather than to put on less and do it more frequently. The larger initial application gave top benefits in the first few years, but the smaller more frequent applications failed to equalize later in the period.
>
> Farmers therefore may lime heavily during a period of prosperity and feel that they have a reserve to carry through a depression. Objections to this line of thought are:

1. That a false sense of security may be created.
2. The period of depression non-liming may be extended over too many years.

Soil testing is the safeguard against this as it enables farmers to base their liming program on known soil reactions at any time rather than on mere guesses.

5. Checking the Lime Level

The progressive farmer will follow the approved practice of checking the lime level of pastures every three or four years by taking soil samples and having them tested by an approved laboratory. The amount of lime required to raise the pH of a soil has been given in previous chapters.

Vary the fertilization program

By varying the fertilization program, the farmer can alter the proportion of legume or grass in the mixture. Grasses love nitrogen; legumes are heavy feeders of phosphate and potash. So, the farmer who wishes to stimulate the grass uses a fertilizer high in nitrogen. On the other hand, if it is the legume that needs stimulation, the farmer omits the nitrogen from the fertilizer and applies phosphate, potash, or both.

On some soils the use of sulfur will stimulate the growth of legumes. The increase in legumes increases the amount of nitrogen in the soil and the subsequent growth of grasses. Thus an evolution is developed within the pasture where one species may crowd out another. The wise use of fertilizers can counteract or influence this tendency and maintain the desired plant mix.

The management of pastures for different classes of livestock is presented in other chapters.

6. Controlling Grazing

One of the major management phases of utilizing productive pastures most efficiently is regulating the time and length of grazing of a particular pasture. This can be done best by rotational grazing, where animals are permitted to graze in one field for a certain time, then are moved to another field. This allows the first field to recover.

According to *The Yearbook of Agriculture: Grass*—

> Controlled grazing is essential for high production of forage crops. All forage crops after clipping or grazing draw upon stored carbohydrate reserves in starting new growth. The reserves are replenished as the plants recover. Various grasses and legumes differ in the extent of depletion of reserves after clipping and in the rate of recovery. Alfalfa is particularly susceptible to continued depletion of root reserves and must not be grazed or clipped too frequently.
>
> The responses of Ladino clover, however, are quite different from those of alfalfa. The stem of Ladino clover grows along the surface of the ground, and a leaf develops at each node. When growing with a tall, vigorous grass, each new leaf of clover arises in rather dense shade; under such conditions, new growth draws heavily on the plant reserves. This weakens the plants and may lead to complete loss of the clover.
>
> Highly significant, however, is the fact that Ladino clover stores carbohydrate reserves in the prostrate stems, but the grasses store a large part of the reserves in the lower part of

Courtesy, USDA

Fig. 7.3—A desirable grass-clover combination.

the stem. Thus close grazing actually will remove part of the reserves of the grass but not of Ladino clover. Frequent grazing to hold the grass in check is essential to maintain Ladino clover in a mixed sod. That is particularly important if the grass has been stimulated by nitrogen fertilization.

Divide pastures

One good practice is to divide a big pasture area into several small areas. In this way, livestock can be rotated among the different fields easily and with a minimum of effort. Another recommended practice is to have available in early spring a high-producing pasture.

Where livestock are allowed to graze over a wide area, some of the plants become tall, woody, and therefore unpalatable, and are never eaten. Other areas, on the other hand, are grazed closely and continuously.

Stimulate early growth

Early growth can be stimulated by a late winter application of high-nitrogen fertilizer or by a fall application of manure.

7. Preventing Severe Grazing

Severe grazing, early in the season, at mid-season, or late in the season, usually is detrimental to the stand. It must be remembered that grass develops virtually a new root system each year. These new roots form from food manufactured in the leaves. If the top is eaten off, roots will starve and the plant will winter-kill.

Permit top growth

Permitting pasture plants to make top growth in the fall is therefore essential to give plants winter hardiness and to start them right the following spring.

When herbage is permitted to make considerable growth in either the spring or fall, or better yet at both times, moderate grazing

Courtesy, USDA

Fig. 7.4—This pasture is properly grazed—closely but not too closely.

throughout the remainder of the season is not injurious. In fact, grazing is necessary to maintain the legume or legumes in the pasture. Moderate grazing, however, does not mean that the animals should abuse the pasture. Sufficient growth to shade and cover the ground is always desirable.

Overgraze annuals

Most of the practices thus far refer to perennial plant species and on conditions where year-round moisture is provided either by rain or by irrigation. In some areas, especially in the arid West, annual species may be the only plants in the pasture mix. These plants die down and regrow each season. Once these annual plants have matured and gone to seed, it is a desirable practice to overgraze the area with livestock. In this process the seed is shattered and trampled into the ground, ready and receptive for growth during next year's rainy season.

8. Clipping Pastures

Systematic mowing is very important to pasture management. Mowing prevents the seeding of many weeds and discourages undesirable plants such as briars.

Likewise, if kept cut close, many weeds will be eaten along with the grass-legume herbage and will tend to gradually disappear from the sod. Mowing also removes unpalatable seed stems and uneaten grass, especially around manure droppings. Often this uneaten grass will be picked up after mowing and eaten by the animals when they crave dry feed. The mower should be set to cut closely. Under humid conditions many pasture specialists recommend at least two clippings, the first in June and the second in July.

Regulate number of mowings

The livestock producer can easily determine the number of mowings needed. He or she should never let the grass get out of hand or become uneven, or let weeds go to seed.

Fig. 7.5—Well constructed and properly located cattle guards can regulate livestock on or off pastures without the irritation of opening and closing gates.

The control of weeds by modern chemical herbicides is discussed in a later chapter.

9. Handling Poisonous Plants in Pastures

There are a great variety of plants toxic or even lethal to animals and in some cases humans. Most livestock fortunately shy away from eating plants poisonous to them unless extremely hungry or through accidental ingestion. Poisonous plants can be found on irrigated and non-irrigated areas or on open range. Complete eradication of poisonous species is the most desirable goal, although, for a variety of reasons, this may not be possible in many areas. Therefore, the use of approved practices is the place to start.

Learn to recognize primary species

While a large number of plants known to be poisonous exist in any given area, generally only a limited number will cause serious losses. With sheep, for example, St. Johnwort, death camas, lupines, and water hemlock are of first importance. The prevalence and concentration of a species will give livestock producers real concern. It is desirable then to learn to recognize and identify the plants in your area and especially on your own ranch.

Eradicate toxic species

Where practical, this is an excellent method of preventing losses. Grubbing out and cutting are frequent methods used. Chemical sprays and occasionally soil sterilants are used in well defined patches. In any event, complete eradication by any means is the goal.

Fence off danger zones

Growth may be so thick, particularly in bogs or swamps, that eradication is too costly. Fencing off may be the only practical solution. In time, favorable plants may crowd out toxic species if no grazing occurs in the meantime.

Drain wet areas

Bogs may favor the growth of undesirable species and prevent growth of good grasses. Therefore, spreading excessive moisture by drainage, especially to drier areas, will improve conditions for desirable species. Leveling land may in some cases be the best procedure, as both areas (wet and dry) will be improved.

Graze with proper livestock

Various kinds of livestock select different species of plants to eat. Furthermore, they may vary in their susceptibility to the poison. For example, cattle are rarely poisoned by death camas, whereas sheep losses are heavy. Sheep, on the other hand, seldom seem to be poisoned by larkspur, yet cattle frequently have serious losses.

Avoid too early seasonal grazing

Many plants are more toxic during one growth stage than another. Due to this fact and also because many poisonous plants make a fairly good growth before beneficial plants, early seasonal grazing is a particularly dangerous time. Some plants may be safe at a particular stage of development in the plant, yet toxic at another period. Yellow star thistle is readily and safely eaten by all classes of livestock, yet horses often will continue to graze on it during the sticker stage until their mouths become so sore that they refuse to eat and starve to death.

Be alert to new poisonous plants

With today's rapid transportation, historically clean areas often can be infected with new toxic plants. Halogeton and Medusa Head are two examples. These plants have spread rapidly into areas that just a few years ago were clean. Much of the problem is associated with rough land where irrigation, cultivation, or spraying is difficult or impossible. If one is alert and recognizes the problem early, control measures can be initiated before it becomes widespread.

CHAPTER 8

The Nutritional Value of Pastures

The livestock farmer should be interested not only in high yields but also in feeding value of herbage grown. The aim should be to produce as much total digestible nutrients per acre as needed, in forms best suited to the livestock enterprise. In previous chapters the effects of management practices on total yields have been discussed. In this chapter we will deal with those factors and practices that influence the nutritional value of the herbage.

Activities Which Involve Approved Practices

1. Selecting plants for best nutritive value.
2. Treating soil to improve feeding value.
3. Fertilizing to increase nutritive value.
4. Obtaining most feeding value from plants.
5. Increasing the profit from pastures.

1. Selecting Plants for Best Nutritive Value

The best pasture mixture for an individual livestock farmer depends on the particular type of livestock enterprise. Soil conditions on the farm, climatic conditions of the area, and the labor available to carry out the farming operation are vital factors.

Generally, a dairy farmer will profit more by growing a pasture mixture that produces a large tonnage of crude protein. On the other hand, a beef farmer pasturing mature animals is not necessarily as much interested in the protein content of the pasture as in total digestible nutrients. The beef animal's requirement for protein is less than that of a cow producing large quantities of milk.

The reader is referred to the chapters on pasture management for individual classes of livestock for specific information.

Fig. 8.1—For mature or growing animals, good pastures provide Nature's most perfect feed.

Legumes are protein-rich

Legumes are especially high in protein and, when included in the pasture mixture, raise the protein content of the herbage. This increase is brought about in two ways. In addition to being high in protein themselves, inoculated legumes actually furnish nitrogen to the grasses in the mixture. This nitrogen, in turn, increases the protein content of the grasses. Legumes are also rich in calcium, phosphorus, and other minerals. Their use in mixtures, therefore, also increases the mineral content of the herbage.

Use deep-rooted legumes

Many of the legumes are deep-rooted and are therefore less sensitive to drought conditions than grasses. The inclusion of one or more legumes in the pasture mixture is usually an approved practice. The management practices usually center around maintaining a desirable grass-legume combination. These practices have already been covered in previous chapters.

2. Treating Soil to Improve Feeding Value

Fertilizers and lime improve the feeding value of the pasture herbage. First, by raising the fertility status of the soil, conditions are made more favorable for the growth of the desirable species (the nu-

Courtesy, W. J. Meade

Fig. 8.2—Undesirable plants cover far too many pasture fields.

tritious legumes and grasses) in a pasture mixture. As a result, a higher percentage of the total yield is made up of plants rich in protein and high in minerals.

The use of lime where needed favors the growth of legumes. A good example of how the use of fertilizer increases the growth of clover and palatable grasses is illustrated by a Virginia Agricultural Experiment Station project in which a field was selected that contained approximately 75 percent broomsedge. One portion of the field was fertilized with a mixed fertilizer supplying 48 pounds of nitrogen, 48 pounds of available phosphate, and 48 pounds of available potash per acre. An adjoining area was left untreated. The fertilizer treatment was repeated every year for four years. In the spring of the fourth year, stand counts showed that bluegrass and white clover had almost crowded out the broomsedge on the fertilized land, leaving only 8 percent of the area still covered with broomsedge. On the non-fertilized section, however, there had been no decrease in broomsedge.

The California Agricultural Experiment Station has found that the same thing happened on one of its red terrace soils. Applications of 112, 225, and 450 pounds of concentrated superphosphate increased the percentage of legumes in the forage from about 45 percent for the non-treated area to 57 percent for the plot receiving the heaviest phosphate fertilization.

3. Fertilizing to Increase Nutritive Value

The second way that fertilizers and lime improve value of pasture is by raising the nutritive value of the individual plants making up the mixture. Usually when small applications of fertilizer are applied, there is an increase in total yield but little or no increase in feeding value per pound. However, with the use of liberal quantities of fertilizer there are generally both an increase in yield and an improvement in the quality of the herbage—an increase in the protein and mineral contents of the individual plants. This is especially true on soils that are relatively infertile.

Change plant analysis with fertilizer

We have already mentioned the California experiment where the use of concentrated superphosphate increased the proportion of legumes in the mixture. In this same experiment, chemical analysis showed that the crude protein content of the whole forage was raised from 16.4 percent to 19.3 percent where the 450-pound-per-acre superphosphate application was made. Part of this increase in protein content came from the larger proportion of high-protein Ladino clover in the mixture, and part from an actual increase in the protein content of the grass.

Other chemical tests revealed that the superphosphate application also increased the phosphorus content of both the grass and the Ladino clover. The analyses showed the following: The phosphorus content of the mixed grasses ran 0.185 percent when no phosphate was applied; 0.253 percent when 112 pounds of superphosphate was used; and 0.390 percent when 450 pounds of superphosphate was used.

The mineral content of the Ladino clover also increased with each additional application of superphosphate. The clover grown on the area receiving no superphosphate carried 0.161 percent phosphorus; with increased use of superphosphate, the phosphorus content of the clover reached 0.244 percent where the 450-pound application was made.

Fig. 8.3—This pasture is too mature for highest feeding value.

At Beltsville, Maryland, the U.S. Department of Agriculture also found that applications of lime and fertilizer increased the phosphorus and calcium content of pasture herbage. Bluegrass in the treated area was 25 percent higher in phosphorus and carried 16 percent more calcium than bluegrass from the non-treated area. There were also increases of 22 percent in the phosphorus and 12 percent in the calcium content of the fertilized clover over that of clover grown without fertilization.

4. Obtaining Most Feeding Value from Plants

Graze at proper growth stage

The feeding value of pasture plants decreases as the plants mature. The farmer is, therefore, faced with controlling the grazing so as to provide a high yield without sacrificing much in the way of quality. If the farmer grazes too early, the total tonnage-yield and, to a

Courtesy, USDA

Fig. 8.4—Rotational grazing is an approved practice that pays off in increased production.

lesser extent, yield of total digestible nutrients will be reduced. If the farmer grazes too late, the plants are woody and unpalatable, and low in feeding value. The stage of grazing best suited for an individual farmer is dependent upon his or her livestock enterprises. Certainly a dairy farmer would graze at an earlier stage of maturity than would a farmer pasturing mature beef cattle.

In the spring, when the new grass is beginning to grow, the water content may be as much as 85 percent. Therefore, the 1,000-pound animal would have to eat 100 pounds of such fresh forage to obtain 15 pounds of dry matter, which is but little more than the requirement for maintenance.

As pasture plants mature, the percentage of protein and minerals decreases. For example, air-dried bromegrass in North Dakota cut on May 10 contained 18.5 percent of crude protein, whereas that cut on July 25 contained only 9.2 percent. Within the same time the ash content decreased from 11.9 percent to 5.7 percent.

At another location, samples of Nevada bluegrass, violet wheatgrass, and Letterman needlegrass from Utah contained in the dry matter 25 percent protein on June 24; 11 percent on August 9; 10 percent on August 29; 6 percent on September 18; and 5 percent on October 7. If such forage is cut about the time the seed matures, the plants contain only about 40 percent of water. With so much greater growth and lower moisture content, the grazing animals can readily get not only enough dry matter for maintenance but also plenty for fattening. Forty pounds of such forage would contain 24 pounds of dry matter, which is practically a full feed for a 1,000-pound animal. If the same forage is cut twice during the season, the water content of the fresh forage would be about 54 percent. When cut four times a season, the average water content is about 79 percent.

Mature forages lose a considerable part of their nutritive value by exposure to wet weather. As the more soluble and digestible nutrients are leached out by rain, the less valuable part for feeding therefore remains. The greatest loss is the soluble mineral matter. Such losses may exceed 60 percent. Consequently, it is very important that a mineral supplement be supplied to livestock fed chiefly on leached roughage.

The effect of the leaching action of rain, accompanied by the loss of leaves, is illustrated by bur clover in California, When it is ripe, cattle prefer it to the green feed that is available and fatten rapidly. But if much rain falls on the ripe clover, it loses its leaves and the feeding value is not even sufficient for maintenance in many cases.

In the arid and semi-arid sections of the West, where the pasture and range grasses cure in their natural state with little or no loss of leaves, these grasses have nearly the same feeding value as hay made from the same plants and ordinarily carry cattle, horses, and sheep through the winter in good condition, if a plentiful supply is available.

5. Increasing the Profit from Pastures

The quality of forage does make a difference to livestock. This point was illustrated by a feeding experiment in California. In this test, steers were fed 91 days. One group of animals was given hay analyzing 0.2 percent phosphorus while another lot received hay containing 0.1 percent phosphorus. The low-phosphorus hay actually

Courtesy, Allis-Chalmers Mfg. Co.

Fig. 8.5—Desert areas are generally poor pastures. However, they can be improved and made useful through practices of fertilization, controlled grazing, and reseeding to improved species and varieties.

Courtesy, American Jersey Cattle Club

Fig. 8.6—Vigorous, high-producing dairy cows and a productive pasture make a profitable team.

graded U.S. No. 1 extra leafy while the high-phosphorus hay graded only U.S. No. 2 leafy. In other words, the low-phosphorus hay appeared to be better.

In the feeding trial, however, the low-phosphorus hay did not come up to the hay with the higher phosphorus content. Where it took 940 pounds of high-phosphorus hay to give 100 pounds of gain, it took twice this amount of hay containing only 0.1 percent phosphorus.

In research tests at the University of Minnesota, scientists J. D. Donker, G. C. Marten, and W. F. Wedin concluded that, from continuous comparisons of different levels of concentrate intake, most Holstein cows producing close to 50 pounds of milk daily could be supported by pasture alone. There were indications that cows over the 50-pound production level would decline unless concentrates were provided. Higher levels of production appear to need supplementation with concentrates.

Work done at the Grand Rapids station of the same university was summarized by George Oraskovich as follows: Phosphate and potash doubled the TDN and milk production over the unfertilized pasture field where 36 pounds of P_2O_5 and K_2O, respectively, per acre were applied to a largely bluegrass pasture, while the addition of 200 pounds of nitrogen per acre with the phosphorus and potassium on another field almost doubled the TDN again.

CHAPTER 9

A 12-Month Forage Program

The dairy, beef, horse, and sheep producer must provide animals with quality forage 12 months out of the year. A sound pasture program, especially in areas where grasses and legumes do not provide year-round grazing, is not adequate for a 12-month forage program.

Most permanent pastures produce nearly two-thirds of their growth during the first third of the growing season. The usual practice of undergrazing in the spring to conserve growth for later is a wasteful practice with most forage species. Kentucky bluegrass in the humid region retains its feed quality quite well, even when in advanced maturity.

Profitable livestock production requires that most of the forage be converted to livestock products. Grasses and legumes are most nutritious before they mature. Light grazing in the spring permits the grasses and legumes to mature beyond their most palatable stages of growth. In addition, the animals also trample considerable forage into the ground. The net result is low production with considerable herbage not being eaten.

The only sensible solution is a 12-month forage plan that provides quality forage the year round. There is no one plan. After considering soils, climate, equipment and storage facilities, the farmer should develop a specific 12-month plan for each farm unit, according to needs.

Annuals and perennials should be used. Corn silage is a high-quality feed, and yield potential is high. On highly productive soils, corn silage usually produces higher yields and more utilizable energy than perennial grasses and legumes.

Farms with a good distribution of moisture and a long growing season can develop programs where livestock are on pastures most of the year. Separate chapters deal with emergency and winter pastures. These pastures, when managed in accordance with approved practices, can in many areas extend the grazing season. These practices should be considered and used if the economics favor their use.

In most of the United States, a sound 12-month forage program involves the utilization of pastures, hay, and, in many instances, silage. Silage may be grass and/or corn silage.

In general, forage production on an individual farm might be increased by use of different plant species to give a longer seasonal production with higher total yield, use of better cultural practices, and use of plants with higher productive potentials. Adoption of systems to reduce waste during grazing, harvesting, and feeding and damage due to overuse or trampling by livestock is another factor to consider. Combinations of the foregoing may be the answer to production of more high-quality pasture and forage.

Several of the suggested approved practices can be recommended only if herds are large, since machinery costs are high. On the other hand, the principles of mechanization in farming have been dictated from the standpoint of economy and have resulted in one person being able to handle a larger number of animals.

Legumes, adapted legume-grass combinations, or high yielding, high-quality grasses must be used where they will perform best on the soils and under the conditions on each farm.

Activities Which Involve Approved Practices

1. Producing high yields of grasses and legumes.
2. Avoiding continuous grazing.
3. Employing standard rotational grazing.
4. Using strip grazing.
5. Utilizing forage at palatable stages of growth.
6. Using green chopped pasture.
7. Using dual-purpose pastures.
8. Feeding silage (haylage).
9. Making quality hay.

1. Producing High Yields of Grasses and Legumes

Regardless of the crop, yields must be high if the unit cost—cost per pound of gain or cost per ton of forage—is low. This is true because of the high fixed costs associated with farming. Taxes and fencing costs are the same whether an acre produces 100 pounds of beef or 600 pounds of beef. If the crop is harvested mechanically, the cost is essentially the same per acre for a 2-ton hay crop as for an 8-ton crop. Regardless of yield, it takes just about as long to mow the field and bale or cube the hay. There is, of course, a bit more cost to handle the extra yield, but it is small when figured on a ton basis.

High yields don't just happen. The use of adapted grasses and legumes is an absolute necessity. The varieties must have the potential to produce good yields of quality forage and then be managed so the potential is realized.

Pastures should be grazed to obtain high animal production per acre. At the same time, grazing should be controlled to maintain quality and quantity for suitable production per animal. A lush growth in the spring with low summer and fall yield does not fulfill the needs of livestock. This situation is especially troublesome for a cow and calf enterprise. With cows and calves (or steers) the animals need more feed as they increase in size during the pasture season. With later summer freshening of dairy cows, feed needs are also highest when pastures are lowest in production.

Because of peaks and valleys in pasture production, there is no 100 percent practical stocking rate that will utilize all the herbage without a surplus and/or deficit during certain periods. Patterns of grazing will go a long way in making full utilization of forage and are therefore discussed in this chapter.

2. Avoiding Continuous Grazing

If pasturing a herd in one field for the season may be called a system, then the system of continuous grazing, generally, is considered poor management. It could mean smaller investment in fence and lower management cost, but it should be used only on dryland pastures where it is impractical to cross-fence or in grazing the aftermath of hay or silage.

Since animals graze herbs they like and avoid others, maintaining the desired composition of grass and legumes is difficult, weed control becomes complex, and the problem of taking advantage of high-yielding grass and legume mixtures is encountered. Species vary during the season in their rate of growth and nutrient composition, so if the carrying capacity of the forage is gauged by mid-season production, spring season growth will go ungrazed and mature to increase in crude fiber and lignin content, with consequent decrease in protein content and palatability. Both overgrazing and undergrazing can occur with such a system. Overgrazing can weaken and kill pasture grasses and legumes, while undergrazing can lessen the ratio of such legumes as Ladino clover and annual lespedezas, since they do not tolerate the competition of undergrazed tall grasses.

3. Employing Standard Rotational Grazing

This system represents a modification of the Hohenheim system long used in Western Europe and consists of confining livestock to an area small enough so they will eat all pasture plants to a desired height in a given period of time. The pasture is divided into several paddocks by use of fences (usually electric), and the livestock are moved every week or so.

Maintenance of desirable pasture species and more efficient use of forage are the important considerations in this system. Rotation grazing reduces damage to forage from trampling, particularly of tall, dense forage crops, since the animals require less time to supply their needs. Since paddocks are alternately grazed and left idle, such practices as fertilization, weed clipping, and irrigation can be carried out more effectively, and excess forage can be harvested more easily for hay or silage. Since plant growth rate varies during the season, flexibility in timing permits movement of the livestock to be slower in the spring during lush growth or relatively faster when the plant growth slows down later in the season. The net result is an overall increase in forage production of 25 to 50 percent over continuous grazing. Desirable composition of high-yielding grasses and legumes can be maintained more easily.

Often, it is desirable to combine with the rotational system provisions for supplying emergency or supplemental forage of extra pasture, harvested forage, or grain. This would be desirable when pasture growth slows down because of drier conditions of summer or the cyclic nature of growth of some plants, or to extend the beginning or ending time of the usual pasture season. Such crops as sudangrass and small grains are examples of crops used as supplementary forages.

4. Using Strip Grazing

This is a system of intensified rotation grazing where the animals are provided smaller grazing areas such that only the amount of forage they will consume in a given time, usually a day, or occasionally one-half day, is available to them. Usually there are fewer fluctuations of animal production, bloat is less likely to occur because animals generally eat some of the more mature forage, and more flexible use of the forage is provided than in standard rotational grazing.

Although there is very little wasted forage in this system and weeds are not a problem, clipping of pasture areas is recommended in order to remove clumps of uneaten forage and to promote young and nutritious forage growth. Hay or silage may be made from any pasture area, if excellent growth conditions cause a surplus for grazing.

5. Utilizing Forage at Palatable Stages of Growth

The quality of feed produced can be controlled in part by timing the harvest, whether this is done by grazing or mowing.

The stage of growth is the single factor most likely to influence chemical composition and feeding value. Immature grasses and legumes are succulent, high in protein and moisture, but low in fiber. As maturity increases, the plants approach the reproductive stage. Stems which produce the flowering parts appear and elongate rapidly. These stems have a high dry matter, high carbohydrate, and a higher crude fiber content than the leaves and a lower crude protein percentage. If the plants are permitted to reach maturity, the increasing percentage of crude fiber decreases both palatability and digestibility.

Active growing forage contains 80 to 85 percent water. Legumes are normally higher in moisture than grasses. In studies at the University of California (Davis) all pasture mixtures increased in dry matter as the forages became older. One mixture of legumes and grasses averaged 16.5 percent dry matter when cut at two-week intervals and 21.4 percent when cut at five-week intervals.

The crude protein content of young forage is usually more than

Courtesy, New Holland Division of Sperry Rand

Fig. 9.1—Wilted hay crop makes good green chop or silage.

adequate. Grass pastures seldom have less than 10 percent crude protein on a dry weight basis. Grass legume pastures usually average 20 percent or more crude protein, and rapidly growing pastures may run as high as 30 percent. The digestibility of this crude protein may range from 50 to 85 percent.

In harvesting, the goal is to get as much high-quality forage as possible. Grazing techniques to achieve this goal include rotational grazing or strip grazing. Rotational and strip grazing have already been discussed.

6. Using Green Chopped Pasture

Green chopping and feeding directly to animals during the growing season will probably provide the most nutrients to a herd without waste. This system represents an adaptation of soiling (green cutting) and feeding forage crops to supplement pastures first used in the United States in the early 1900's. Interest in the system was revived with the advent of higher-yielding forage species and such labor-saving machinery as field choppers, self-unloading wagons, and automatic feed distributing equipment.

Feed intake can be controlled as to quantity and quality, and permanent fencing practically eliminated. Briefly, this system enables the manager to meet the forage needs with fewer acres than would be required by grazing, because waste from selective grazing, trampling, and fouling of forage by animals is avoided. Furthermore, both the leaves and stems are harvested and fed. The need for fencing, watering facilities, shade, and other requirements of grazing systems are eliminated.

In the case of a dairy herd, for instance, at least 60 cows are required to make green feeding economical, and a high degree of crop management skill is involved in successful operation of the system. Supplementing green feed with silage solves the problem when wet weather makes fields impassable for equipment and results in no loss in production since the livestock are already accustomed to silage conditions.

7. Using Dual-Purpose Pastures

Green forage for early spring and fall grazing and for hay during winter is a forage management system tried in the shortgrass plains area. Native grasses are largely warm-season species and are not adapted to hay production. Intermediate wheatgrass has been found to be a late-maturing species which begins growth early in the spring and which makes good recovery from grazing if moisture is available in late summer and fall. The grass is palatable to sheep and gives fairly good hay production. Russian wildrye has been found to be a drought-resistant bunchgrass adapted to the Great Plains regions, readily grazed by sheep at any time of the year, but due to a mostly basal foliage characteristic, not suitable for hay production. Pubescent wheatgrass has been found to be drought-resistant and palatable to livestock and to have good hay qualities. Crested wheatgrass has been found to be winter-hardy and drought-resistant and to make rapid early spring growth for spring grazing, and good growth when fall moisture is available.

Ladak alfalfa has been found to be adapted to the Great Plains region, produces a good first crop, recovers slowly, but gives generally good growth by October 1.

Further information can be obtained from a University of Wyoming publication, "Dual-Purpose Pastures for the Shortgrass Plains."

A five-year study of the economics of pasture systems of 54 south-

Courtesy, New Holland Division of Sperry Rand

Fig. 9.2—Balers with bale throwers reduce labor in hay harvesting.

ern Michigan dairy farms by C. R. Hoglund and C. M. Harrison led to some interesting observations. Forage crops accounted for 60 percent of the crops grown, with alfalfa-bromegrass being the major crop combination grown for summer feed as well as for hay for the barn period. The number of cows averaged about 33 for the farmers using field grazing and strip grazing, 38 for those green-chopping, and 51 for those feeding out of storage. Farmers who used strip grazing or green chopping made greater use of sudangrass for mid-season feed. The cows in all four groups were fed supplementary hay and silage during the pasture season. It was concluded that strip grazing would be most profitable for farms with less than 30 cows and that most dairy farmers would benefit most by making improvements in the forage practices and pasture systems they were using.

8. Feeding Silage (Haylage)

The silage (haylage) system can be a completely mechanized one. In field operations, the forage harvester can be coupled with the self-unloading wagon and with a forage blower or elevator to deposit the material efficiently in an upright silo. For trench or bunker silos, the use of a self-unloading wagon simplifies matters. Unloading operations can be accomplished mechanically by use of mechanized silage unloaders on tower silos, tractor front-end loaders in trench or bunker silos, and mechanically driven belt-paddle arrangements for distribution to feed bunks.

With this system, silage is available throughout the year and is under complete control of the manager. If combined with the green feeding system, the filling of the silo from the first cutting can proceed rapidly to take advantage of the high digestibility values of the forage and permit the recovery growth to get off to an early start. Bottom unloading, airtight silos lend themselves well to this system of forage handling. Climatic conditions do not seriously interfere with silage-making operations.

Haylage

A distinction should be made between grass silage normally stored unwilted and with a moisture percentage of around 75-85, and haylage (low-moisture silage) where the plants are wilted after being cut and allowed to dry to a moisture percentage of 40 to 60. Such forage can be stored only if air is completely excluded; therefore only those types of silos that can be sealed are used for haylage preservation. There would be too much risk of spoilage in trench silos. Proper handling of low-moisture silage, or haylage, eliminates odors, requires no preservatives, provides more protein and vitamin D in the feed, and encounters less freezing trouble during the winter.

Wilted silage

Wilted silage is a little higher in moisture than haylage, but drier than direct-cut grass silage. Wilted silage is usually cut with a windrower-conditioner and allowed to sun dry for four to eight hours. The drying time is just long enough for the moisture to reach 65 percent. This moisture has been found to be the most ideal for ensiling legume-grass forages. Minimum storage and field losses are realized at this moisture.

With an all-silage system, a farm will require more silo capacity

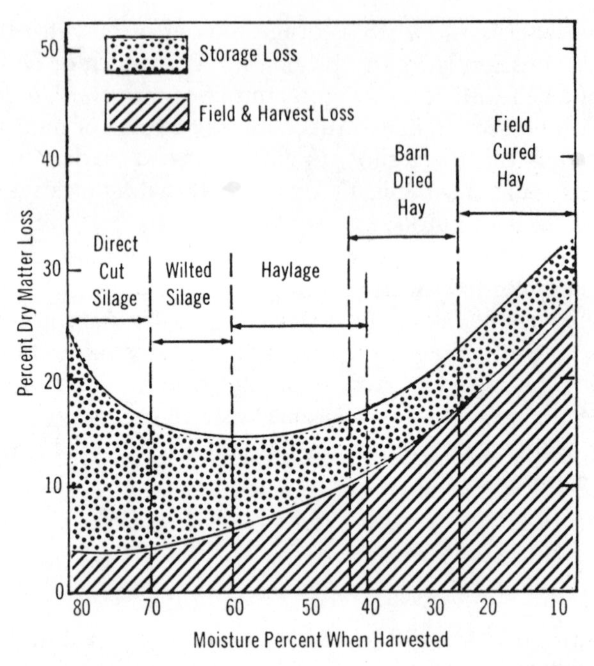

Courtesy, Michigan State University

Fig. 9.3—Estimated total field and harvest loss, and storage loss when legume-grass forages are harvested at varying moisture levels and by alternative harvesting methods.

than where the system is combined with other forage handling systems.

Chopped hay is made from forage that has been mowed and allowed to wilt to lower the moisture content to 30 percent or less.

After this, the material may be picked up by the forage chopper, conveyed to a wagon, and transported to a barn designed to store, cure, and allow self-feeding of hay. The barn designed to store high-moisture chopped hay has ducts through which heated or unheated air is forced to dry the hay. The hay may be baled by a pick-up baler but not before the moisture content is down around 22 percent, unless the storage facilities are provided with drying equipment. The hay may be pelleted or wafered, usually after being dehydrated or put through a hammer mill. Long hay has been made into a cube suitable for dairy cows. Machines have been developed capable of making pellets from windrowed hay in arid regions of the West where the moisture of the hay may get down to 10 percent. Supplemental moisture is added at the pelleting machine to make a hardened surface which will resist physical abuse. This allows the pellets to be handled pneumatically, augered, or moved by drag type conveyors.

9. Making Quality Hay

Most grass-legume mixtures, if harvested early enough, make good-quality hay. Generally, this hay is cut and baled. If it is high in legumes, such as alfalfa, it is sometimes cubed in regions where field-dried hay can get down to 10 percent moisture.

Hay-making of surplus forage or of forage grown for this purpose is an excellent way of conserving forage for the livestock for periods when pastures are dormant or are not producing adequately to meet the animals' needs.

As with pastures, fields harvested for hay should be cut before the plants become woody. Usually, the stage of harvest is delayed somewhat later than when pastured to increase total yield and facilitate

Courtesy, Hesston Corporation

Fig. 9.4—Modern hay-making equipment cuts labor requirements.

Table 9.1

Early First Harvest Saves TDN and Crude Protein in Alfalfa-Grass Mixtures

Stage of Growth and Date of First Cutting	Percent TDN	Percent Crude Protein	Percent Fiber
Pre-bud (immature), June 2	70.8	20.1	23.9
Late bud (1/10 bloom), June 15	62.3	17.2	30.4
Half bloom (maturing), June 24	59.3	14.3	32.5
Full bloom (mature), July 2	56.0	12.5	34.8

Table 9.2

Cutting Frequency and Harvest Date Determine Forage Quality

Treatment Number and Cutting Dates	Pounds TDN per Acre	Pounds Crude Protein per Acre	Tons Dry Matter per Acre
I. Early—3 cuts on June 2, July 15, and August 31	4,152	1,192	3.2
II. Median—3 cuts on June 15, July 27, and August 31	4,470	1,273	3.5
III. Late—2 cuts on June 24 and August 15	4,041	1,013	3.4
IV. Very late—2 cuts on July 2 and August 31	4,056	986	3.7

Table 9.3

Alfalfa-Grass Quality of First Cutting Changes Rapidly

Component	Change per Day	Total for 31 Days
TDN	−0.48	−14.8
Protein	−0.25	−7.6
Fiber	+0.35	+10.9

curing and handling. The harvest should be timed and handled to conserve the leaves. Harvesting should be done while the plants are still green and all care possible exercised to conserve the green color.

Alfalfa is a common legume in many pasture mixtures. Smooth bromegrass, orchardgrass, or timothy commonly will be found in mixture with alfalfa. These combinations make good yields of excellent hay. There are varieties of alfalfa that are adapted to essentially all areas of the country. This legume has the potential of increasing yields wherever the soil and climatic conditions are favorable. Where successfully incorporated into the mixture, alfalfa increases production whether the field is grazed, green chopped, or harvested for hay.

Since alfalfa by itself or in mixtures is a plant widely used for hay, data are presented showing the effect of stage of harvest on yield and quality, the effect of cutting frequency and harvest date on quality, and the effect of delaying harvest on quality. These data are presented in Tables 9.1, 9.2, and 9.3 and are from research conducted by the University of Minnesota.

CHAPTER 10

Pastures for Dairy Cattle

The modern dairy cow has been bred for the efficient conversion of large quantities of forage into milk. It has been given this description by *The Modern Dairy Cow:*

> The cow was designed by the Almighty for the consumption of forage crops. This animal is a mobile harvesting machine, equipped with a mower and grinder on one end and a fertilizer spreader on the other. In between these two extremes is located an extremely complex manufacturing plant designed for the conversion of large quantities of raw materials—pasture, silage, hay—into Nature's most perfect food—milk.

How the Dairy Cow Uses Feed Nutrients

The dairy cow requires nutrients for:

1. Body development and maintenance.
2. Milk production.
3. Reproduction.

Milk production is more frequently limited by an insufficient supply of feed nutrients than by any other cause.

If a cow produces 9,000 pounds of milk, this means she has eaten less than 11,000 pounds of feed with about 55 percent digestibility. This is only 30 pounds of air-dried feed per day. Researchers in New York kept records on a number of New York dairy farms. Averaging the data on 175 cows for 20 weeks starting 60 days after freshening, they found that cows would eat an average of 26.7 pounds of forage dry matter per day along with 9.6 pounds of grain. The range in feed intake of forage dry matter was from 12.9 to 35.8 pounds; grain intake ranged from 3.8 to 16.8 pounds. The production averaged 39.6

pounds FCM (fat converted milk) ranging from 18.2 to 65.2 pounds. Using these figures as averages, 36.3 pounds of feed produced 39.6 pounds of 4 percent FCM.

The point to be made is this: High-producing cows should not be fed forage alone. It takes grain along with forage to manufacture production energy. Forage will provide the maintenance energy and most of the protein needed to balance the deficit of feed grains. A good dairy producer should recognize the fact that 1,000 pounds of forage dry matter feed can produce more than 1,000 pounds of milk provided this forage is fed as a part of a balanced ration. The only feed supplement needed with high-quality forage is grain and minerals. A cow will not eat much pasture if she can find her protein and fiber in the manger.

Activities Which Insure Profits from Pastures

1. Determining feed requirements to meet goals for the herd.
2. Determining what is needed for supplemental feed.
3. Obtaining the most economical source of supplemental feed.
4. Providing the proper levels of supplemental feed.
5. Using rotational grazing to prevent ill effects of mismanagement and to insure optimum pasture consumption.
6. Clipping when necessary to prevent seedhead development in grasses.
7. Spreading nitrogen fertilizer after clipping grass and dragging to spread animal manure droppings.
8. Using tough permanent sods such as tall fescue for loafing areas, night pastures, and rainy weather pastures.
9. Providing effective shades to reduce fly problems and ill effects of heat.

1. Determining Feed Requirements to Meet Goals for the Herd

Well-managed pastures provide a nearly perfect feed. However, an extremely lush pasture may lack energy. With such a feed-saving potential, a major emphasis should be given to pastures to stretch production to at least 8 months and in some regions 11 or 12 months of the year.

A good pasture system should yield 5 to 6 tons per acre, of which 4 tons would be consumed. This 8,000 pounds of dry matter should provide the maintenance energy and protein for at least three cows. Such a feed package should also account for 8,000 pounds of milk, according to current feed standards.

2. Determining What Is Needed for Supplemental Feed

The Virginia Agricultural Experiment Station has conducted a great deal of research involving supplemental feeds used in conjunction with pastures and hay. One experiment was planned to measure milk production with (1) pasture only, (2) pasture during night hours with hay in a shaded dry lot during the day, and (3) hay in dry lot day and night. Three rates of concentrate supplements were also investigated. The pasture was mainly orchardgrass–Ladino clover, but small grains were grazed during the early spring and alfalfa-orchardgrass during part of the summer. Excellent quality alfalfa-orchardgrass hay was fed at 15 percent excess consumption.

The average of milk produced daily during two seasons for all concentrate feeds was 16 percent more for pastures than for quality hay. Without concentrates the cows on pasture produced 42.2 pounds of milk daily, 35 percent more than for hay. Cows fed the low concentrate rate (1 pound to 8 pounds milk) produced considerably more milk on pasture than on hay. With liberal concentrates, milk production per cow for pastures and hay was similar.

The researchers concluded: "Concentrate supplements increased milk production for all pasture and hay treatments because the energy content of forages, especially hay, was inadequate." Partial data from this experiment are presented in Table 10.1.

3. Obtaining the Most Economical Source of Supplemental Feed

It has been established that good pastures—as good as they are—do not provide sufficient energy to permit high-yielding cows to produce maximum milk. These pastures do provide the needed protein.

Table 10.1

Milk Produced Daily per Cow with Three Rates of Concentrates[1,2]

	Milk per Cow at Three Rates of Concentrates		
	None	**Low**	**Medium**
1. Milk produced daily during a pregrazing standardization period of 2 to 3 weeks when concentrates were fed at the rate of 1 lb. to 4 of milk	40.7 lbs.	42.0 lbs.	42.9 lbs.
2. Rates of concentrates—lb.:lb. of milk[3]	none	1 : 9	1 : 45
Rates based on consumption	none	1 : 93	1 : 5.1
3. Milk produced per cow—average	35.0 lbs.	38.5 lbs.	42.2 lbs.
4. Percent of original (standardization)	85%	91%	92%

[1]*Source:* Virginia Agricultural Experiment Station.
[2]Milk not corrected to 4% fat. Produced during three grazing seasons on orchardgrass–Ladino clover pastures.
[3]The grazing seasons were 187, 133, and 118 days, respectively, for three consecutive years. The concentrate feeds for each rate are based (milk per cow) during the pregrazing standardization period and were kept constant.

Therefore, the supplemental feed should be looked upon as a source of energy. The grains such as corn and milo are excellent sources of energy and have a high palatability rating. The judicious dairy farmer will shop around for energy-producing feed so as to purchase what is needed at the lowest cost. In some instances, it may be a by-product feed. The farmer's concern is one of getting energy with palatability. The use of good mineral supplements is also an approved practice since the mineral requirements of high-producing cows are high.

4. Providing the Proper Levels of Supplemental Feed

The exact quantity of supplement a dairy farmer should use will depend on (1) the production potential of the cows, (2) the quality of the pastures of forage, and (3) the supplement:milk ratio cost. Data

Table 10.2

Decreases in Dry Matter Digestibility Lower Forage Intake and Milk Production and Increase Grain Requirement

Cutting Date	Dry Matter Digestibility	Digestible Forage Consumed	Milk Produced	Grain Required for Maximum Production
	(%)	*(lbs./day/ 1,000 lbs. of body weight)*	*(lbs./day)*	
5/17	66.8	23.0	42.5	3.0
5/24	65.0	21.6	39.2	5.7
5/31	63.1	20.2	34.0	8.4
6/7	61.3	18.8	31.4	10.9
6/14	59.4	17.4	26.5	13.5
6/21	57.5	16.0	23.4	15.7
6/28	55.8	14.7	19.5	18.2

[1]*Source:* Ohio Agricultural Experiment Station.

collected at the Ohio Agricultural Experiment Station show that a decrease in quality of forage increases grain intake. See Table 10.2.

These data reveal that dry matter digestibility decreases at about 0.3 percent per day after May 12, or 2.1 percent per week. Not only does digestibility decline but also the amount consumed daily is reduced. Cutting date (for a given species) is the best index to forage digestibility for an on-the-farm situation estimate. Digestibility dropped 11 percentage units or 16.5 percent, but milk flow dropped 54 percent due to the drop in digestible forage consumed. The grain required for maximum production increased sixfold.

5. Using Rotational Grazing to Prevent Ill Effects of Mismanagement and to Insure Optimum Pasture Consumption

It is false economy to turn the average-sized dairy herd into a large field of improved pasture. When this is done, the herd wanders over the field eating a bite here and a bite there. Much of the grass is trampled down and is never eaten. The sensible approach is to turn

Courtesy, USDA

Fig. 10.1—Good pastures, like the one at the right, are "money makers"; poor pastures, like the one at the left, are "money losers."

the animals into a small area, leave them there until they have eaten down the herbage, and then move them to another area.

Graze for short periods

An approved practice is to use areas that the herd will graze down in four to six days. Then the herd is moved onto another ungrazed area; the cows are not rotated back on the first area until the stand has recovered and has made considerable regrowth. This approved practice greatly increases the carrying capacity of the pasture.

Improved pastures handled in this manner produce large quantities of milk.

Fig. 10.2—Electric fences make rotational grazing practical and profitable. They should not be used as boundary fences or along busy roads.

Rotate grazing to help herd maintain production

Tests at the Virginia Agricultural Experiment Station showed that rotating and grazing pastures for a period of four to six days at a time resulted in greater persistency of milk production by the herd. In other areas, progressive dairy farmers rotate their herds to a new grazing area each day. They usually use an electric fence, moving it daily to fence in sufficient new pasture growth to furnish adequate feed but little wastage for their herds.

Haul in forage (green chopping)

In intensive farming areas, other farmers follow the practice of cutting the herbage at its most nutritious stage with a field chopper, and then hauling and feeding it to the herd in a dry feed lot. This system of feeding permits the full utilization of all the herbage since there is no waste from trampling and manure droppings. When coupled with high fertilization and irrigation where needed, tremendous acre yields are possible. This system of pasture utilization requires less nutrients for body maintenance since the animal does not use up energy in gathering her feed.

6. Clipping When Necessary to Prevent Seedhead Development in Grasses

Even under rotational grazing, the herd will not graze the entire area uniformly. There will be areas where the cows have lain down, where droppings have fallen, and where, for other reasons, the herbage has not been eaten down. If the pasture is to be kept growing vigorously and evenly, it is necessary that all the vegetation be of the same stage of maturity. So any herbage not consumed by the animals should be mowed off just as soon as the herd is moved from the pasture. When clipping is not practiced, the grasses are likely to crowd out the legumes in the areas not grazed down by the animals.

Follow with dry cows

Another approved practice which works well for some farmers involves pasturing the milking herd until a considerable portion of the lush herbage has been consumed and then following the herd with dry cows and calves. This practice is called "top grazing." High-production animals graze off the tops—the most digestible part of the pasture plants, and dry cows and heifers are called "second grazers," getting what's left.

Regardless of the management, any uneaten herbage should be clipped as soon as the cows are moved to another area. When this is done, new growth bounces back quickly and the grass-legume mixture is easier to maintain.

7. Spreading Nitrogen Fertilizer After Clipping Grass and Dragging to Spread Animal Manure Droppings

Grasses have very healthy appetites for nitrogen and potash. If these nutrients become limiting, yield and quality drop materially.

Many dairy producers, making the most of their pastures, spread fertilizer high in nitrogen after each clipping. This readily available nitrogen results in a quick, succulent vegetative growth that is high in protein. The higher the legume content, the higher the potash requirement and the lower the nitrogen need since legumes can "fix" their own nitrogen.

Researchers at the University of Minnesota took a pasture consisting of 85 percent bluegrass with the remainder being quackgrass, timothy, and bromegrass. They applied 40 pounds of phosphate (P_2O_5) and 50 pounds of potash (K_2O) per acre. The first 50 pounds of nitrogen was applied early in the spring and then 50 pounds after each field was grazed until the top rate was reached. Table 10.3 gives the results in cow days per acre, milk per acre, and carrying capacity. The 200-pound nitrogen rate produced threefold that of the no-nitrogen treatment field.

Table 10.3

Nitrogen Applications Increase Milk Production[1]

Pounds Nitrogen per Acre	Cow Days per Acre	Pounds Milk per Acre	Cows per Acre for 150-Day Season
0	135	3,178	0.42
50	246	6,121	0.84
100	352	8,564	1.14
200	424	10,705	1.38
300	346	8,864	1.18

[1]*Source:* University of Minnesota.

The Minnesota scientists, in commenting on the results of this and other research findings, make this statement:

> Fuse to the powder keg is recognizing that bluegrass doesn't burn up—but that it runs out of food reserves. Feed it and it grows, and grows, and grows. High summer temperatures and small amounts of rainfall impose limitations. But even these can be partially overcome. Grass fertilized in early spring develops a thick sod. This heavy sod and the foliage above it insulates the growing parts of the grass plant from hot summer temperatures. It also holds moisture and leads early rainfall down into the soil for storage.

Soils research has shown that split applications of nitrogen automatically help farmers hedge against periods of hot weather and drouth. If grass does not produce enough for grazing, then no more nitrogen is applied.

Scatter droppings

Manure droppings should be scattered by using a harrow or similar implement. Although mowing helps dissipate the droppings, normally it does not spread them uniformly over the field. Unless the droppings are scattered, the grass growing around them is not relished by the livestock. There are indications that this grass, although high in protein, is low in sugars.

Eliminate odors

There are others who reason that the livestock do not graze such areas because of the odor associated with the droppings.

Regardless of the cause, the fact remains that milk production per acre will be cut unless the manure is scattered. The scattered droppings also return valuable plant food nutrients to the soil.

8. Using Tough Permanent Sods Such as Tall Fescue for Loafing Areas, Night Pastures, and Rainy Weather Pastures

Constant hoof tramping will kill out or destroy most grasses and legumes. Tall fescue, where adapted, can stand the punishment better than most other species. This grass forms very tough mats of roots that can carry heavy traffic. Tall fescue grows well early in the season and stays green late into the fall. It is a vigorous grower which recovers rapidly after grazing.

A good sod not only provides feed but also keeps the cows clean and reduces soil erosion. Heavy, dense, quick-recovering-type grasses are ideal for areas that are heavily stocked, such as holding pastures next to the barns.

Some farmers also make it a practice of having the loafing area of permanent sod, such as tall fescue or bluegrass, close to their improved

Fig. 10.3—A stile, strategically placed, eases the task of getting in and out of fenced pastures.

pastures. Where this is done and where such areas are shaded, the animals usually will go into the improved pasture, graze their fill, and then return to the shaded area to chew their cuds and make their milk. This practice, of course, eliminates herbage wastage since the animals will not be lying down in the pasture field and many of their droppings will be confined to the loafing area. Where loafing areas are provided, it is advisable to provide water at the same locations.

Whether a loafing area is provided or not, the importance of shade and adequate clean, fresh water cannot be overemphasized. Both are essential for top milk production.

Let the cows do the work

One of the reasons pastures are so profitable lies in the fact that the livestock do their own harvesting. It takes far less time to feed and care for the cows than when they are barn-fed.

The Ohio Agricultural Experiment Station has found that it takes 12 percent less labor to do the daily chores when the dairy herd is on pasture than when it is barn-fed. There is nothing wrong about that important labor-saving figure of 12 percent. Even the time required taking the cows to and from the field was taken into account.

9. Providing Effective Shades to Reduce Fly Problems and Ill Effects of Heat

Missouri dairy specialists talk about cow discomfort in terms of *temperature-humidity index.*

The good dairy farmer will use shades, shelters, fans, sprays, and other means to keep cows comfortable. These devices can go a long way in reducing the fly problems and the ill effects that heat may cause.

CHAPTER 11

Pastures for Beef Cattle

Beef cattle lead all classes of livestock in the consumption of pasture grasses. They utilize about one-third of the permanent pastures and three-fourths of the range area. The average beef animal in the humid areas eats approximately 50 pounds of green herbage daily. In the drier sections of the country, the daily intake of the herbage is about 25 pounds. The dry matter intake being approximately the same, the difference in weight is explained by the variation in moisture content.

S. E. Clarke and associates of the Canadian Department of Agriculture find that about 25 pounds of native grass of southern Alberta, Saskatchewan, and Manitoba are required to make a pound of beef and leave sufficient pasture growth as carryover. The grasses of southern Canada are very nutritious and contain less moisture than most of the grasses in humid areas of the United States. However, on the basis of the Canadian studies, a 1,000-pound steer fed nothing but grass from weaning time would have consumed approximately 25,000 pounds of pasture herbage.

The management of pastures to give the greatest dollar return per acre, year in and year out, is the number one consideration of the beef producer.

Activities Which Involve Approved Practices

1. Getting the right pasture mixture.
2. Managing for high carrying capacity.
3. Extending the grazing season.
4. Handling bloat.
5. Practicing rotational grazing.
6. Grazing native rangeland.

1. Getting the Right Pasture Mixture

In experiments conducted over a four-year period by the Florida Agricultural Experiment Station, yearling steers on an improved white clover–carpetgrass pasture produced 619 pounds of beef per acre as compared with 219 pounds for steers on a carpetgrass-lespedeza pasture.

Fig. 11.1—Beef cattle thrive on good pasture.

In the Northwest, researchers found that grass-alfalfa mixtures produced approximately three times as much beef as pure grass pastures of smooth bromegrass or crested wheatgrass. The alfalfa-brome mixture gave 225 pounds of beef per acre in 87 days; the pure stand crested wheatgrass, 90 pounds of beef in 85 days.

Essentially all of the states' agricultural experiment stations conduct experiments to determine desirable pasture mixtures. The reader should write to the station in his or her state for recommended mixtures suited to the area.

2. Managing for High Carrying Capacity

The beef producers are in the business to make a profit. The producer is merely using cattle to harvest forage and sell the forage product. It is essential, therefore, that the total yields of the forage be good and that the quality be high. Forage not harvested, but wasted, returns no profit.

The goal of successful beef producers is to provide adequate nutritious feed to their animals every day and at the same time avoid waste that may result during "spurt" growth periods. In other words, they attempt to stock their pastures to match production. Surplus forage is harvested for feeding during slow growing periods.

Dr. N. D. Morgan has calculated data which show how carrying capacity, which of course is a measure of management, influences returns. Dr. Morgan made these calculations from actual farm situations and research data from the University of Arkansas, Louisiana State University, Oklahoma State University, and Texas A&M University. These data are presented in Table 11.1.

Table 11.1

It Takes High Fertilizer Rates to Make Money Regardless of the Prices (on a 100-Acre Livestock Farm)

	This Just Happened (Average Management)	This Was Planned (Recommended Management)
Stocking rate of 100 acres	25 brood cows	100 brood cows
Fertilizer cost	Little or none	$ 2,700
Total cost and investment (land, cattle, equipment, etc.)	$35,206	$70,796
Profit (for labor management)		
With good market $35	$ 3,500[1]	$ 9,000[2]
With poor market $25	$ 23.25	$ 4,000

Low yields caused by low soil fertility and high soil acidity will result in low profits and more farmers losing their farms.

[1]When interest of $2,500 on land and cattle is deducted there is only $1,000 ($10 per acre) left for land and management.

[2]When interest on land, cattle, and equipment is deducted, return to labor and management is $5,040 or $50 per acre.

The key to success is to produce high-quality forage and use that forage while it is young and tender. All unused forage results in potential profit being left in the field. "Produce and utilize" could well be the slogan of all who pasture beef animals.

3. Extending the Grazing Season

In most areas it is possible to produce a variety of pastures that will provide good early grazing, mid-season grazing, and late grazing. As an illustration, Dr. Walter Wedin and R. L. Vetter of Iowa State University have developed a "three-season" grazing system that can increase beef production sixfold over unimproved pasture.

In Monroe County, Iowa, the management system of heavy fertilization and rotation grazing produced 600 pounds of beef per acre.

Pure stands of tall fescue, reed canarygrass, smooth bromegrass, and orchardgrass received 240 pounds of nitrogen per acre in three applications. This included 40 pounds per acre in March or April, 80 pounds per acre at the end of May, and 120 pounds per acre early in August. In addition, a 13-pound-per-acre application of phosphorus and a 50-pound-per-acre application of potassium were applied either in the fall or in the spring.

Since the work was experimental, the researchers do not offer these rates as a recommendation. However, this combination certainly produces beef.

Each grass species was grown by itself in a pure stand without a legume. Each grass plot was divided by an electric fence into two-thirds and one-third. A water tank was placed between the areas.

The larger area received cattle about May 1. The cattle were rotated to the smaller area three weeks later. (One area can be cut for hay if enough cattle are not available to graze the forage.) The area was then clipped, and 80 pounds of nitrogen fertilizer per acre was applied.

At least two grazing rotations were alternated on these areas. Easy surface application was achieved by the application of granular ammonium nitrate. During the last two or three weeks of July the entire area was grazed, and the cattle were then removed from the plot. Pastures were clipped and fertilized with 120 pounds per acre of nitrogen.

In August and September and the first half of October, the pastures were allowed to rest. About the middle of October, cattle were

Fig. 11.2—Large-frame sires improve herd performance.

once more placed on the one-third area for two or three weeks, then the entire pasture.

Such a management pattern breaks the grazing into six or seven periods of three or four weeks each. Wedin and Vetter call all periods "three seasons"—spring, summer, and fall.

These findings stem from research aimed at determining how much beef could be produced per acre with controlled management of both grass and animals, according to Wedin.

In the past, researchers have been able to stretch out the spring grazing season by about 10 days. This was accomplished by fertilizing. Capitalization on the lush fall growth, however, has not been attempted. It is thought that increasing the fall grazing period could have a twofold effect. One effect would be extra grazing, and another would be a substantial reduction in the amount of stored feed needed.

"Greatest daily gains surprisingly enough, have been obtained in the fall period," explains Wedin. "However, spring and summer gains are also good," he continues.

Even in cold weather, cattle seem to do well in the fall, though the grasses have ceased to grow. The cattle feed on the lush growth that has accumulated before the freezing.

Table 11.2 illustrates gains for each species used. Tall fescue is not too palatable in spring and summer, but it provides good carrying capacity because of high yield. This grass would appear best for fall grazing. The highest average daily gain of 3.94 pounds was obtained with orchardgrass at early spring grazing period 1.

Reed canarygrass is best in mid-summer, but not too good in the fall. Considering all periods of grazing, smooth bromegrass was best of the four grasses used. At all seasons of growth, the cattle found it the most palatable. Orchardgrass was a close second. Both of these grasses had good carrying capacities as demonstrated by days of grazing and pounds of beef produced.

"To achieve 600 pounds of beef per acre you have to divide up the grazing season," points out Wedin. "Grazing the tall grasses in August and September is not the thing to do. Use some other kind of pasture, perhaps sorghum-sudangrass, alfalfa-grass aftermath or birdsfoot trefoil."

Table 11.2

Average Daily Gain, Animal Days per Acre, and Pounds of Beef Produced per Acre

	Period	Tall Fescue	Reed Canarygrass	Smooth Bromegrass	Orchard-grass
Average Daily Gain	1	3.01	3.84	3.48	3.94
	2	1.40	2.26	1.06	0.78
	3	1.00	1.12	1.58	0.88
	4	−0.17	−0.03	0.95	0.50
	5	1.85	1.98	2.37	2.11
	6	2.05	1.26	1.76	2.15
	7	2.26	0.71	0.48	0.60
Days of Grazing/Year	1	76	76	84	82
	2	78	52	60	78
	3	74	54	54	50
	4	35	38	27	34
	5	52	47	40	42
	6	64	54	42	50
	7	32	39	30	20
Pounds of Beef/Acre	1	232	287	304	315
	2	99	109	62	54
	3	75	63	84	48
	4	−8	−3	25	17
	5	92	91	88	84
	6	134	73	75	97
	7	72	28	14	12

Wedin also stresses, "To maximize returns from an acre of grass, graze it hard in May and June and again in late fall."

This system may not be so productive as one moves north with the same species of grass. Smooth brome and reed canarygrass would probably be satisfactory there because they are very winter hardy. These species have a spreading growth habit and tend to thicken when properly fertilized.

Anyone going into a high-management pasture program should realize the importance of using his or her own forage. If pastures are undergrazed, money will be lost.

In order to use all forage produced, Wedin put up to 200 head of cattle on 16 acres for a period of two weeks in May. May is the time of greatest growth, and it is important to be prepared to handle this period. Most farmers plan to take part of the forage off as hay in a regular program.

4. Handling Bloat

The grazing of alfalfa or clover in pure stands has not been popular in the United States because of bloat hazard. Mixtures of alfalfa, clovers, and grasses are, however, our most productive pastures. Most livestock producers experience very little trouble with bloat when 50 percent or more of the herbage is composed of the grasses in the mixture.

Bloat is the distension of the rumen caused by excessive accumulation of gases during fermentation of foodstuffs. Death may occur if the gases are produced faster than they can escape from the rumen. The most noticeable symptom of bloat is the distended paunch, particularly on the left side. The respiration is quickened and the animal appears restless and uneasy.

General Precautions

A number of practices should be employed to prevent bloat from occurring. Provide, at all times, access to some coarse feed such as straw, oat hay, or anything to assist in rumen action. Never turn hungry animals into a lush field; let them fill up on safe feed first so they won't overeat in the new pasture. Keep a watchful eye and remove them from the pasture if it looks like a number are bloating. Weather conditions, such as wind, dew, etc., seem to play a part in causing bloat. Bloat-reducing feed additives have proven satisfactory for many pasture conditions.

The wise cattle producer watches the animals closely and, if indications of bloat are evident, follows one or more of the suggested approved practices.

1. Stand the animal with its head up a slope to relieve pressure.
2. Place a round stick about 1 inch in diameter crossways in the mouth of the animal and tie with a string behind the ears. Vigorous kneading or massaging of extending flank will aid in eliminating gases.
3. In mild cases drench with ½ pint of fresh warm milk.
4. Drench with as much as 12 ounces of raw linseed oil or with ½ ounce of formalin in 1 pint of water.
5. Keep the animal moving to aid in eliminating gases.
6. Where immediate treatment is necessary, a trocar may be used. The puncture is made on the left side about 3 inches below the loin and halfway between the last rib and hipbone. Sterilize instrument to prevent infection. This should be used as a last resort.

Courtesy, USDA

Fig. 11.3—Good pastures and quality livestock are a profitable team.

7. In severe cases, call a veterinarian.
8. Eliminate from the herd chronic bloaters.

5. Practicing Rotational Grazing

At the Indiana Agricultural Experiment Station, a birdsfoot trefoil–bluegrass pasture under rotational grazing produced on the average 250 pounds of beef per year over a two-year period. In the same experiment, an alfalfa-timothy mixture produced 225 pounds of beef, and a Ladino clover–bromegrass pasture gave 190 pounds of beef per acre per year. All three of these mixtures yielded more than permanent bluegrass. The latter produced only 160 pounds per acre.

Moreover, the first three pastures provided more uniform grazing throughout the season than bluegrass, which was productive in both the early and late season but provided little pasture during midsummer.

Match cattle to carrying capacity

In a 10-year grazing study at the South Dakota Agricultural Experiment Station, it was found that beef cattle on a heavily grazed pasture lost weight and went into the winter in a thin condition, whereas animals on lightly grazed pasture gained weight. Under heavy grazing the higher-yeilding, more palatable grasses disappear and are replaced by the shorter grasses, resulting in a decline in range conditions, forage production, and carrying capacity.

The University of Maryland has tested five pasture mixtures at two rates of fertilization. The influence of fertilization and pasture combinations is shown graphically in Fig. 11.4.

6. Grazing Native Rangeland

Graze moderately

Moderation in grazing ranges has shown outstanding advantages in all degrees of grazing tests conducted throughout the West. Generally, total gain per acre has favored the heavier rate of stocking, but

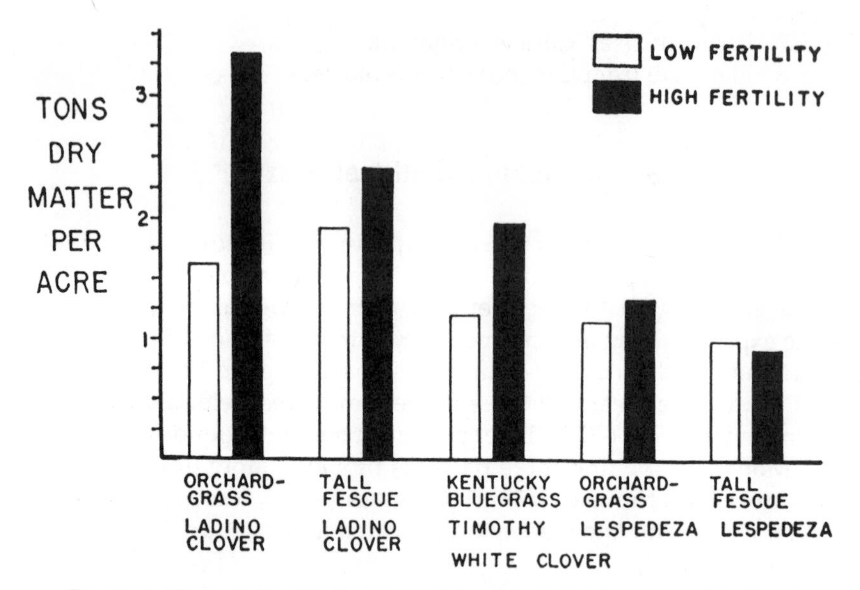

Fig. 11.4—The results of five pasture mixtures tested at two rates of fertilization. Based on the grazing period April 23 through October 5.

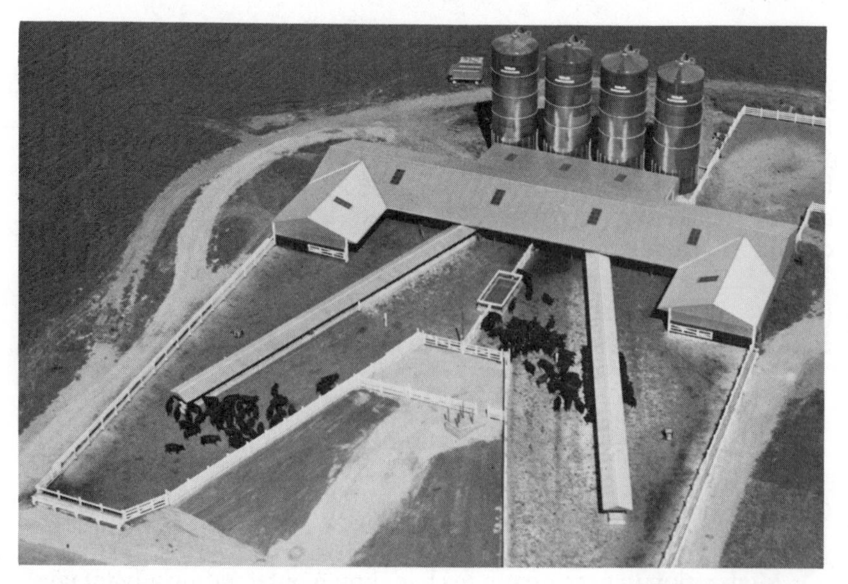

Courtesy, A. O. Smith Co.

Fig. 11.5—These cattle were raised on pasture and are now being finished in the feed lot—the final activity in beef production.

the fallacy of heavy grazing has been consistently demonstrated in reduced gains per head, lower market value of cattle produced, and a decline in the range productivity.

THE "UNBEATABLE TEAM"

The combination of high-yielding grasses and legumes growing on fertile soils can't be beaten. Such combinations are consistent high producers throughout the season. The best combination varies somewhat from area to area. Also, the soil's need for fertilizer varies as indicated in previous chapters.

CHAPTER 12

Pastures for Sheep

Sheep are natural-born grazers and are efficient utilizers of grass. It is estimated that approximately 80 percent of the average sheep's diet comes from pasture and practically all the rest from harvested forage.

Some 70 percent of the sheep in the United States are raised west of the 100th meridian. Therefore, in the discussion of sheep pastures, rangelands cannot be disregarded.

The carrying capacity of most rangelands is far below their true potentials simply because of overgrazing and other mismanagement. The possibilities of increasing the production of sheep by employing approved practices are unlimited.

Activities Which Involve Approved Practices

1. Pasturing rangelands.
2. Pasturing humid areas.
3. Extending the grazing season.
4. Increasing the carrying capacity.

1. Pasturing Rangelands

C. E. Holscher and D. A. Spencer have classified the ranges into 10 broad types. They differ in grazing capacity because of the amount and composition of the vegetation, variations in rainfall, and productivity of the soil. These investigators list the grazing capacities of the various types as follows:

1. Tall grass, 0.5 acre per sheep per month.

Fig. 12.1—Sheep are natural-born grazers.

Courtesy, USDA

Fig. 12.2—Most rangelands are overgrazed.

2. Short grass, 0.8 acre per sheep per month.
3. Pacific bunchgrass, 0.9 acre per sheep per month.
4. Semi-desert grass, 1.1 acres per sheep per month.
5. Sagebrush grass, 1.8 acres per sheep per month.
6. Southern desert shrub, 2.3 acres per sheep per month.
7. Salt desert shrub, 3.6 acres per sheep per month.
8. Pinyon-juniper, 1.7 acres per sheep per month.
9. Woodland chaparral, 2.0 acres per sheep per month.
10. Open forest, 1.6 acres per sheep per month.

2. Pasturing Humid Areas

In the humid areas, sheep are grazed on permanent, rotation, and temporary pastures. The permanent pastures are often confined to the steep, rough lands not suited to cultivation. Meadows, stubblefields, and other fields suitable for producing annual forage crops fit ideally into the grazing program as temporary pastures. Tempo-

Courtesy, USDA

Fig. 12.3—Kentucky bluegrass makes good early spring pasture in the cool humid regions of the U.S.

rary pastures are usually used in periods of drought when the permanent pastures are at a low ebb.

In the North, Kentucky bluegrass makes up a large portion of the vegetation on the permanent pastures. Further south bermuda and other grasses become more important. Each year, there is more interest in the development of pastures for sheep. These pastures are composed of high-producing grasses and legumes. Orchardgrass, bromegrass, alfalfa, Ladino, and white clover are good examples. The clovers and alfalfa provide an abundance of palatable herbage. Care must be exercised in grazing them because of the bloat problem. Mixtures of clovers and grasses are much safer to graze than pure stands of legumes. Because of the increased protein content, legume-grass mixtures are preferred to straight grass, especially for lambs.

3. Extending the Grazing Season

The grazing season can be extended by pasturing small grains both in the fall and in the early spring. In fact, in the warmer areas, small grain pastures can be grazed all winter. Winter rye, winter barley, and winter oats are suited for grazing by sheep. In the more northerly area, rye is preferred because of its winter hardiness and ability to make good growth under cool conditions. Further south, oats and barley produce an abundance of highly nutritious herbage. Some farmers sow a combination of small grains in preference to straight seedings.

The following is adapted from *Minnesota Farm and Home Science*, Vol. XVII, No. 2, pp. 10, 11, and 19.

The success of annual (temporary) forage crops for pasture in emergencies has varied. Yields and protein returns of several of these crops have been reported. However, the actual animal output has been in most cases omitted or only estimated.

To test forages for fattening spring lambs, experiments were started at the Agricultural Experiment Station, Rosemount, Minnesota, using both small plot tests and grazed pastures.

Crops tested in small plots were divided into two groups.

Early summer crops with rate of seeding per acre were:

1. Oats, 3 bushels.
2. Peas, 2.5 bushels.
3. Oats, 1.5 bushels; peas, 1.5 bushels.

4. Oats, 3 bushels; rape, 6 pounds.
5. Oats, 1.5 bushels; peas, 1.5 bushels; rape, 6 pounds.
6. Rye, 3 bushels.

Late summer crops were:

1. Sudangrass, 42 pounds.
2. Soybeans, 2.5 bushels.
3. Sudangrass, 42 pounds; soybeans, 2 bushels; "solid seeded" corn, 2.25 bushels (3 treatments).
4. Field corn (mixed maturities).
5. Sweet corn.
6. Flint × dent corn.

All plots and grazed pastures were located on Waukegan silt loam, well-fertilized on the basis of soil tests. Additional nitrogen was applied before planting.

The average total yields in tons of dry matter per acre for the early summer crops were: oats-rape, 2.35; oats, 2.06; oats-peas-rape, 1.91; oats-peas, 1.46; peas, 1.15; rye, 1.10. For the late summer crops yields were: sudangrass-soybeans, 3.20; corn (maturity mix), 3.14; sudangrass, 3.08; corn (flint × dent), 2.81; corn (sweet), 2.59; soybeans, 2.22. In general, crops that permitted a second cutting yielded more (rape, sudangrass). Rye sown in the spring was poor both in recovery after cutting and in resistance to rusts.

Protein percentages for early summer crops were: peas, 22.8; rye, 21.7; oats-peas, 18.3; oats-peas-rape, 16.6; oats, 16.0; oats-rape, 15.6. Late summer crop averages were: soybeans, 19.5; field corn, 16.0 sudangrass-soybeans, 15.8; sweet corn, 15.2; flint × dent corn, 14.9; and sudangrass, 13.4.

The annual forages and their stocking rates per acre were:

1. Oats-rape pasture continuously grazed throughout the summer (12-18 lambs).
2. Peas grazed until July 1 (24-30 lambs) followed by "solid-seeded" corn (24-36 lambs).
3. Oats-peas grazed until July 1 (24-30 lambs) followed by sudangrass (18-30 lambs).

Thus, the average stocking rate for the grazing season by treatment was in the range of 12-15 lambs per acre.

The researchers concluded:

1. Total yields were highest for forage combinations including at

Fig. 12.4—A fine flock of Dorset ewes. Note that the pasture is divided for rotational grazing.

least one species such as rape or sudangrass with recovery capacity after clipping or grazing.

2. "Solid-seeded" corn was high-yielding but lacked recovery after grazing or clipping. To be used advantageously, a heavy grazing pressure is needed.

3. Protein percentages varied among the crops tested in small plots, with peas, rye, oats-peas, soybeans, field corn, and sudan-soybeans the highest.

4. Average daily lamb gain for the entire grazing period was not affected by the combination of crops used.

5. Oats-rape provided more lamb days of grazing per acre and total lamb produced per acre than either the combination of peas followed by corn or oats-peas followed by sudangrass. In addition, oats-rape provided pasture over a longer period of time.

6. Sudangrass matures rapidly and becomes rank, causing "selective grazing" for lambs.

7. Spring seeded annual forages do not provide adequate forage until about June 1.

8. It may be hard to get a good stand of rape each year unless the pasture is seeded properly.

9. Annual forages should be regarded as temporary or emergency crops and used to augment or complement perennial pastures following winter killing or during periods of low production (hot weather, drought).

4. Increasing the Carrying Capacity

The carrying capacity of different pastures varies with fertility, length of growing season, rainfall, and management practices.

Graze according to pasture

The Wyoming Agricultural Experiment Station has found that the grazing capacity of pastures seeded to standard crested wheatgrass was 143 sheep days per acre. Good native range had a grazing capacity of only 60 days; but when the native range was pitted with an eccentric disc, the carrying capacity was increased to 75 days per acre. Crested wheatgrass produced 89 pounds of lamb per acre, pitted native range 56 pounds, and non-treated native range 29 pounds.

In tests at Beltsville, Maryland, the U.S. Department of Agriculture found that an acre of bluegrass pasture provided 821 ewe days of grazing and a mixture of barley and winter oats, 348 ewe days.

Other temporary pastures included in the experiment were wheat and rye, wheat, rye, barley, and oats. All produced less grazing than the barley–winter oats combination.

CHAPTER 13

Pastures for Poultry

Today, the majority of poultry in the United States, whether for meat or eggs, is produced in confinement. In most instances, confienment rearing achieves sufficient savings in labor, land, and deceased mortality, which more than compensates for the saving in feed that can be had from pasture.

However, since under certain circumstances there can be a savings in mash and grain by use of good pasture, and since a great many turkey producers still use pastures, and since there are still some farm flocks, it seems appropriate to devote a chapter to "Pastures for Poultry."

The percentage of concentrates in the poultry ration that can be replaced with pasture varies. Since plants differ widely in palatability, the species composing a pasture, the volume of forage available, and management practices all influence the utilization of pasture in poultry production.

H. R. Bird, formerly U.S. Department of Agriculture poultry nutrition specialist, and Professor of Poultry Science at The University of Wisconsin, states that it is probably safe to estimate that, under favorable conditions, 10 percent of the dry matter of the diet of chickens can be furnished by good pasture.

Activities Which Involve Approved Practices

1. Saving poultry feed costs.
2. Cutting laying flock feed costs.
3. Using pastures for breeding flocks.
4. Foraging with turkeys.
5. Knowing types of poultry pastures.

6. Practicing rotational grazing.
7. Poultry pointers.

1. Saving Poultry Feed Costs

The Vermont Agricultural Experiment Station reported a saving in feed costs of $51 for each acre of pasture in the growing of pullets as compared with the feed costs of similar birds raised in confinement. In this experiment, 500 Plymouth Rock pullets were grown from 12 to 24 weeks of age on bluegrass, white cover, and birdsfoot trefoil pasture. The birds reared on the range were of better quality than those in confinement. Some social groups are opposed to raising poultry in confinement.

Use Ladino clover

The Ohio Agricultural Experiment Station found Ladino clover pasture preferable to bluegrass. The findings showed that pullets grew 11 percent faster on 12 percent less feed on Ladino clover than similar Leghorn pullets on bluegrass range. Rhode Island Red pullets grew 9 percent faster on 12 percent less feed on Ladino clover pasture than similar pullets on the average farm range.

Save on mash

Other research findings show varying results as to the mash and grain saved by raising birds from 10 to 12 weeks of age to maturity on pasture. Generally, however, the findings substantiate these of H. R. Bird, i.e., that a saving of approximately 10 percent in feed costs and more uniform, better-quality birds can be raised on good pasture than where pullets are raised in confined quarters.

2. Cutting Laying Flock Feed Costs

For the laying flock, producing market eggs only, H. R. Bird estimates that a good pasture can supply 12 percent of the protein, 5 percent of the calcium, 100 percent of vitamins A and D, and 100 percent of the riboflavin necessary to sustain good egg production.

Ohio research workers report a saving amounting to $100 per acre in feed costs for 90 Rhode Island Red hens from April 27 to November 15 in comparison to feed costs of their 90 winter pen mates kept in confinement during the same period.

The hens remaining in confinement received a complete practical laying mash while the hens moved to Ladino clover received a ration consisting of 95 percent coarsely ground corn, 3 percent steamed bone meal, 1 percent salt, 1 percent chick-size oyster shell or limestone, and 1 percent insoluble grit.

Unfortunately, the savings in feed is not sufficient to compensate for the extra labor costs when the flock is large. The modern profit-

Table 13.1

Comparative Growing Mashes[1] for Chicks on Pasture and in Confinement (10 Weeks to Maturity)[2]

Ingredient	Confinement (lbs.)	Good Pasture and Sunshine (lbs.)
Ground yellow corn	563.2	759.2
Heavy oats	200	200
Wheat std. midds.	300	350
Wheat bran	200	100
Soybean meal (solv.)	400	400
Meat scrap	100	100
Alfalfa meal (dhy.)	100	—
Dried whey	20	—
Fermentation solubles (250 mcg. riboflavin/gm.)	20	—
Ground limestone or oyster shell	35	35
Deflu. rock phosphate (34% Ca, 14.5% P)	40	40
Iodized salt	15	15
Manganese sulfate, 65%	0.8	0.8
Vitamin A feeding oil (3,000 A)	4	—
"D" activated animal sterols (2,000 A)	2	—

[1]To be fed with scratch grains; increase the amount gradually until equal parts of grain and mash are fed at 15 weeks of age. Also supply clean water and oyster shell or other calcium grit free-choice.

[2]*Source:* Maryland Agricultural Extension Service Bulletin 95 (Revised).

able poultry venture handles thousands of birds, and the confinement with today's exact feeding including vitamins and minerals makes for lower unit costs with consistent quality of product.

Check market for colored yolks

Birds on pasture produce eggs with deep-colored yolks, and on some markets such eggs bring lower prices than those with light-colored yolks. Where such price difficulties exist, poultry producers may frown on using pasture for the laying flock. However, moving the layers to the range enables the producer to use the laying house for brooding purposes.

3. Using Pastures for Breeding Flocks

For a laying flock producing eggs for hatching purposes, H. R. Bird has estimated that good pastures can supply 12 percent of the protein, 5 percent of the calcium, 12 percent of the magnesium, 75 percent of the riboflavin, and 100 percent of vitamins A and D required to sustain good hatchability.

The ration in Table 13.2 recommended by C. F. Combs of the University of Maryland indicates the ingredients that may be either reduced or omitted entirely for breeding hens on a good pasture.

4. Foraging with Turkeys

Graze with turkeys

Turkeys can utilize more forage than chickens, especially during the later stages of their development. Moreover, turkeys are better foragers than chickens and can get more of their total feed requirements from pastures, provided the volume of forage they can economically consume is available through a systematic pasture program. Land grazed by turkeys has its fertility greatly increased and works well into a rotation planting.

Table 13.2

Comparative Mashes[1] for Breeder Chickens on Pasture and in Confinement

Ingredient	Confinement (lbs.)	Good Pasture and Sunshine (lbs.)
Ground yellow corn	475.5	745.5
Wheat bran	200	100
Wheat std. midds.	300	350
Pulverized heavy oats	200	200
Soybean meal (solv.)	350	350
Fish meal	50	50
Meat scrap	120	100
Alfalfa meal (dhy.)	150	—
Dried whey	50	—
Fermentation solubles (250 mcg. riboflavin/gm.)	20	5
Deflu. rock phosphate (34% Ca, 14.5% P)	58	58
Ground limestone or oyster shell	20	20
Iodized salt	20	20
Manganese sulfate, 65%	0.5	0.5
Vitamin A feeding oil (3,000 A)	3	—
"D" activated animal sterols (2,000 A)	2	—
Vitamin B_{12} feeding supplement (3.5 gms./lb. or equiv.)	1	1

[1]To be fed with approximately equal parts of scratch grains. Supply fresh water and oyster shell or other calcium grit free-choice.

Rotate to different fields

On the other side of the ledger, turkeys, being heavier than chickens, soon destroy the forage they do not consume by trampling unless the poultry farmer practices rotational grazing to preserve the stand.

Where spring and summer pastures are supplemented with a combination such as Ladino clover adjoining a field of corn and soybeans that is planted for grain and ready for harvest by the birds about September 5, it is possible that such a combination of forages-

Courtesy, *Kentucky Farmer*

Fig. 13.1—Turkeys can utilize more forage than chickens.

grain-pasture might well supply up to one-third of the total feed requirements of birds raised for the Thanksgiving market.

Turkeys on good pasture are apparently able to get about the same percentage of protein and calcium as young chickens and all their vitamin A and riboflavin requirements too; sunshine will supply enough vitamin D.

The turkey breeding and growing rations (Table 13.3) show the savings on ingredients for turkeys on good pasture in comparison with the known requirements for breeding birds and growing birds in confinement.

5. Knowing Types of Poultry Pastures

According to Bulletin 95 of the Maryland Agricultural Extension Service, poultry pastures may be divided into:

1. Permanent.
2. Semi-permanent.
3. Annual or temporary.

A bluegrass–white clover combination would be an example of a permanent pasture, as would Ladino-timothy.

Semi-permanent pastures usually provide grazing for three to five years under favorable management and rainfall. They are frequently composed of perennial and biennial legumes and grasses, with Ladino clover generally being the most important species.

Table 13.3

Comparative Mashes for Turkeys on Pastures and in Confinement

	Breeder Mashes[1]		Growing Mashes[2]	
Ingredient	**Birds in Confinement (lbs.)**	**Good Pasture and Sunshine (lbs.)**	**Birds in Confinement (lbs.)**	**Good Pasture and Sunshine (lbs.)**
Ground yellow corn	562.2	824.7	569.0	846.0
Wheat std. midds.	200.0	300.0	200.0	300.0
Wheat bran	200.0	100.0	200.0	100.0
Pulverized heavy oats	250.0	250.0	200.0	200.0
Soybean meal (44% protein)	220.0	220.0	340.0	340.0
Fish meal	100.0	100.0	—	—
Meat scrap	100.0	100.0	100.0	100.0
Dried brewers' yeast	40.0	—	40.0	—
Dried whey	50.0	—	50.0	—
Butyl fermentation solubles (250 gamma riboflavin per gm.)	20.0	7.5	30.0	—
Dehydrated alfalfa meal	150.0	—	150.0	—
Bone meal (steamed)	60.0	60.0	75.0	75.0
Ground limestone or oyster shell	20.0	20.0	20.0	20.0
Iodized salt	15.0	15.0	15.0	15.0
Manganese sulfate	0.75	0.75	1.0	1.0
Vitamin A feeding oil (4,000 A)[3]	6.0	—	3.0	—
"D" activated animal sterol (1,500 I.C. units D/gm.)	4.0	—	4.0	—
Choline chloride (25% mix)	—	—	2.0	2.0
Vitamin B_{12} supp. (3.5 gms./lb. or equiv.	2.0	2.0	1.0	1.0

[1]Approximately equal parts grain and mash. Oyster shell or other calcium supplement should be fed free-choice.

[2]Scratch grain should be fed free-choice starting at 12 weeks. Oyster shell or other sources of calcium should also be supplied at that time.

[3]Omit at least one month before marketing if a fish oil is used, in order to avoid fish flavor.

Annual or temporary pastures, such as small grain crops, can be used advantageously under certain conditions.

Permanent pastures

Permanent pastures have the advantage that once established they last for a long time. Rotational grazing usually is required to keep them productive. The continued accumulation of droppings year after year contaminates the soil and favors infestation by the common internal parasites of poultry. Roundworms, tapeworms, coccidiosis, and other forms often infest permanent pastures used regularly by poultry.

Under such conditions, it is virtually impossible to raise a healthy flock on such pastures.

Semi-permanent pastures

Semi-permanent pastures when grazed rotationally are probably the most satisfactory under average conditions. They furnish an abundance of palatable, nutritious forage, and since the pastures are re-established every three to five years, the danger of soil contamination is much less of a problem than with permanent pastures. Where there is sufficient land, one of the best ways to grow semi-permanent pastures is in a rotation with corn and small grains.

Practice three-year rotation

Three years of semi-permanent pasture provide abundant grazing and a good supply of grain and bedding, and aid in keeping the range free of parasites since the land is free of poultry the two years it is in corn and small grain.

Temporary pastures

Temporary pastures are the most effective type for breaking the life cycle of parasitic organisms. The frequent cultivation necessary to maintain annual pastures exposes the organisms to sunlight, which de-

Table 13-4

Relative Merits of Temporary Pasture Plants for Poultry[1]

Plant	Palatability	Nutritive Value[2]	Length of Growing Season	Volume of Forage Supplied
Cereal Grains:				
Barley[3]	Poor	Excellent	Short	Fair
Oats[3]	Excellent	Poor to fair	Short	Fair
Wheat[3]	Good	Excellent	Short	Fair
Rye[3]	Very poor	Excellent	Fair	Fair
Sorghum[4]	Fair to poor	Very poor	Long	Good
Young corn[4]	Fair	Very poor	Long	Good
Legumes:				
Soybeans (hay)[4]	Fair to poor	Good	Mod. long	Good
Cowpeas[4]	Fair to poor	Poor	Mod. long	Good
Crimson clover[3]	Fair	Poor	Mod. long	Fair
Vetch[5]	Fair to poor	Fair to good	Short	Fair
Lespedeza	Fair	Fair to good	Poor	Good
Others:				
Rape[4]	Fair	Poor	Long	Good

[1]*Source:* Maryland Agricultural Extension Bulletin 95 (Revised).
[2]Based on data of Morrison for nutritive value.
[3]For late winter and spring pasture.
[4]For summer and early fall.
[5]For spring or fall pasture.

stroys them. However, the additional expense involved in the annual preparation and seeding puts these pastures at a disadvantage over the permanent and semi-permanent types.

In selecting the combination of pasture plants to be used, the poultry producer should keep in mind that it is generally to his advantage to obtain the greatest possible feed supply from this source.

The principles involved in establishing poultry pastures are identical with those already discussed in Chapter 5.

6. Practicing Rotational Grazing

A systematic plan of rotational pasturing has very definite advantages. By rotational grazing is meant the moving of feeders, roosts, and waterers in systematic steps at intervals from one side of the pasture to the other and repeating these steps on permanent and semi-permanent type pastures. The advantages of rotational grazing are:

1. The forage is close at hand for the birds.
2. The moving avoids bare spots through killing of plants by overgrazing and trampling of certain spots, especially those near feeders and waterers.
3. Moving the birds to fresh ground aids in sanitation and helps control diseases.
4. The process of cleaning and handling manure is largely eliminated, as rotational grazing spreads the droppings on the land.

7. Poultry Pointers

Chickens and turkeys graze primarily in early morning when the forage is crisp and succulent. It is important that the birds be released from shelter not later than sunrise if they are to utilize pastures fully.

Provide daily attention

Regular daily attention to feeding and watering and care of the birds on the range is very important to their proper growth and production. Adequate water, shelter from storms, protection from thieves and predators, and the prompt removal of dead birds are important details in making poultry pastures pay best.

CHAPTER 14

Pastures for Swine

Although many hogs are being raised today in confinement, the use of pastures is still in practice in some areas.

Research findings have demonstrated that the concentrate requirements for a breeding herd can be reduced 75 percent for sows and 20 percent for pigs. Fattening pigs on pasture requires 15 percent less concentrates.

Activities Which Involve Approved Practices

1. Obtaining good gains.
2. Knowing essentials of good hog pastures.
3. Knowing types of swine pastures.
4. Self-balancing rations.
5. Preserving sod.

1. Obtaining Good Gains

Pigs on quality pasture can make excellent gains and may reach market size at a lower cost per pound than pigs fed balanced rations in dry lot. Good legume pastures furnish an abundance of minerals and vitamins.

In addition, the pasture can replace about one-half the protein supplement ordinarily fed in dry lot operation.

Sanitation is much easier when pigs are on good pastures, and the danger of parasitic infestation is greatly reduced.

At the Irrigation Experiment Station, Prosser, Washington, fattening pigs going on pasture at 70 pounds of weight returned from $40 to $135 per acre in terms of concentrates saved. Daily gains of fattening pigs in dry lot were 1.75 pounds per day; on pasture plus corn, 1.48 pounds per day; and on pasture plus corn and protein, 1.60 pounds per day.

2. Knowing Essentials of Good Hog Pastures

The U.S. Department of Agriculture lists the essentials of good pasture crops for swine production as follows:

1. They must be adapted to the soil and climatic conditions under which they are grown.
2. They must be grown at small expense.
3. They must provide abundant growth for a short pasture period, or consistent growth over a long period.
4. They must be palatable and succulent so as to be readily consumed and rich in protein, vitamin, and mineral nutrients.
5. They must have a good carrying capacity over the grazing period.

3. Knowing Types of Swine Pastures

Permanent, rotation, and temporary pastures all have a place in swine production. Many successful farmers use all three types. They work out their pasture programs to provide maximum grazing, with planned rest periods for all pastures. The object of their programs is to have the hogs on each pasture when the animals can get the most nutrients without handicapping the carrying capacity of the pasture later on.

Alfalfa and clover are the best legumes for grazing by hogs. They are rich in protein and high in nutrients, and furnish an abundance of herbage over long periods.

Sweet clover also is often used for swine pasture. Best results are obtained when this legume is grazed during the first crop year. The growth during the second year is coarse and less palatable than that of the previous year.

Permanent pastures

Permanent pastures that are used successfully include Kentucky bluegrass, white clover, bermudagrass, carpetgrass, and dallisgrass. These plants provide the most nutrients during the spring and fall and have a dormant period during the hot summer months.

Rotation pastures

Rotation pastures include alfalfa, red clover, and sweet clover. These have already been discussed in a preceding paragraph. Other plants commonly used are Ladino clover, orchardgrass, bromegrass, and lespedeza.

Mixtures of legumes are used extensively in many areas. Combinations provide succulent herbage at almost any season, when the right combinations are chosen. Ladino clover, a high protein yielder, is almost invariably used in the mixtures along with other plants such as alfalfa, alsike clover, bromegrass, and orchardgrass.

Temporary pastures

Temporary pastures are often used during periods when permanent or rotation pastures are not providing sufficient nutritious herbage for most economical gains. Small grains, like oats, rye, or combinations of small grains and winter legumes (oats and field peas are examples) are valuable for late fall, winter, or early spring grazing.

Soybeans can furnish an abundance of excellent green forage in mid-summer.

Rape is another plant that is used extensively by hog growers. Although not a legume, this plant compares favorably with alfalfa and clover. It has a long grazing season and a high carrying capacity. The Dwarf Essex variety is the more palatable to hogs. It may be planted any time during the growing season. It grows quickly and can be planted in combination with small grains or field peas.

4. Self-balancing Rations

Many growers have their sows farrow their litters on pasture. Others farrow in central houses and move the sows and pigs to clean

Courtesy, USDA

Fig. 14.1—Sanitation is easier on pastures. The incidence of disease is reduced and esthetics improved.

pastures at an early age. Either method simplifies feeding and stimulates the sow's milk flow.

Provide separate ingredients

The most satisfactory way of balancing the diet of hogs on pasture is to let them have access to grain, protein supplement, and mineral mixture in separate compartments in a self-feeder.

5. Preserving Sod

"Ringing," the placement of a metal ring through the fleshy tip of a hog's nose, is recommended for hogs on pasture to keep them from rooting holes in the sod. The ringing operation should, of course, be carefully done to eliminate possibility of infection. Usually hogs given a well balanced ration fortified with adequate minerals are less likely to root than those on a mineral-deficient diet.

CHAPTER 15

Pastures for Horses

Although the draft horse has just about disappeared from the farm scene, there still remains a great interest in the breeds of horses used for pleasure and sporting events.

Pastures furnish ponies and other horses with high-quality, nutritious feed at a relatively low cost and help to maintain healthy animals by furnishing exercise, sunshine, and fresh air.

Mounts, mature horses, and younger horses (perhaps well-developed yearlings) can be maintained on good pasture with little or no grain during a normal pasture season. Young animals, those doing medium to heavy work, or those being prepared for shows and contests need grain and hay along with pasture.

No other one feed is as complete a ration for the horse as good pasture. Mineral-rich pasture herbage is considered a necessity in the diet of the horse. Grass from productive pastures furnishes the vitamins and minerals required by the horse. The herbage also satisfies the need for bulk.

Horses that have access to green forage regularly have fewer ailments than those confined to dry feed. Even though pasture is available only for a very short time, it is considered a conditioner and tonic. Young, tender grass from fertile fields seems to have properties beyond those capable of determination by chemical tests.

Green herbage is rich in carotene, which the animal in turn converts to vitamin A. The minimum daily vitamin A requirement of the mature horse has been estimated at about 2.0 to 2.4 micrograms per pound of body weight. This amount can easily be supplied by 1½ pounds of fresh young pasture grass.

If there is a vitamin A deficiency, certain symptoms develop such as a rough coat, failure to reproduce, respiratory difficulties, and night blindness. Unless the deficiency is corrected, death follows.

Courtesy, USDA

Fig. 15.1—Good pasture provides a near-complete ration for horses. Horses can paw through snow to grass and survive whereas cattle cannot.

Activities Which Involve Approved Practices

1. Selecting the pasture mix.
2. Establishing and maintaining the stand.
3. Managing the pasture.

1. Selecting the Pasture Mix

The first decision to make is whether the pasture is to be used for exercise or to provide a major source of energy, protein, minerals, and vitamins. The principles of establishment are essentially the same for either purpose. The differences are mainly in kinds of plants and the amount of land required per animal for the two purposes.

When acreage is very limited (less than 1 acre per horse), exercise may be the main use of your pasture. Pasture for this purpose, in addition to providing exercise, offers many other benefits, including access to fresh air and sunlight and freedom from respiratory problems frequently associated with stabled animals. However, this type of pasture cannot be expected to supply more than a minimum amount of feed.

Well-limed and -fertilized Kentucky bluegrass in areas of adaptation should be the main grass for this type of area. It withstands close and continuous grazing better than most other grasses, and when well established and properly fertilized, it produces a reasonably dense and attractive sod.

In heavy traffic areas, along fences, and around gates and water troughs, tall fescue may be used. While it is generally considered less palatable than bluegrass, tall fescue produces one of the toughest, heaviest traffic sods of any of our adapted grasses.

Grass is the horse's natural food. Productive, well-managed pasture can provide most of the forage requirements for horses during the growing season, including protein, vitamins, and minerals—and usually at a lower cost to you.

On the other hand, poorly managed, worn-out pastures not only supply little or no feed, but are frequently the source of many internal parasites found in horses.

2. Establishing and Maintaining the Stand

Old, worn-out pastures can also be improved by complete renovation, i.e., destroying the existing species and establishing productive mixtures. This procedure usually results in the highest yield increase per acre, but also may be relatively expensive to perform.

If one plans to renovate an old pasture, consideration should be given to the following points:

1. Correct any poorly drained areas either by use of drain tile or by construction of any necessary conservation facilities. Pastures on wet soils are difficult to establish and maintain for sharp-hoofed, fast-moving horses.
2. Test the soil for lime and fertilizer requirements. This is the only sure way of knowing how much lime and fertilizer is needed.
3. Apply required lime or other soil amendments prior to any tillage operation. Discing or plowing will help to mix them evenly throughout the soil. Remember, lime moves slowly through the soil, and it takes several months for lime to neutralize acid soil and to assure efficient use of fertilizer elements. Thus, if possible, lime should be applied and worked in several months before the actual seeding.

4. Destroy the old sod by plowing or by discing. The ideal approach would be to plow or disc heavily in the late summer or early fall followed by seeding the following spring.

5. Complete the job in early spring by adding the required fertilizer and seeding to a good mixture of improved species.

Soils vary in depth and internal drainage. While most of our improved pasture species are useable for horse pasture, the mixtures in Table 15.1 appear to be suitable, especially in the Midwest.

Table 15.1

Seeding Rates for Horse Pastures

Soil Drainage	Seeding Mixture	Seeding Rate (lbs./acre)
Well drained	Kentucky bluegrass	6
	Timothy	4
	Ladino clover	1
	Red clover or alfalfa	4
Varying drainage	Kentucky bluegrass	8
	Smooth bromegrass	6
	or	
	Kentucky bluegrass	6
	Smooth bromegrass	4
	Empire birdsfoot trefoil	6
Poorly drained	Reed canarygrass	8
	Empire birdsfoot trefoil	6

6. Use the band or broadcast seeding method either with or without a companion crop. Where no companion crop is used, annual broadleaf weeds can be controlled with a timely application of a recommended herbicide.

7. Protect the seeded area until the seeded species are well established. It is often assumed that several years are required before a newly seeded pasture is suitable for grazing by horses. However, where good mixtures are seeded without a companion crop and weeds are controlled chemically, good turf can be developed in a single year.

3. Managing the Pasture

Avoid overgrazing or undergrazing. Since horses are notorious spot grazers, they will seriously damage desired species in some areas unless they are frequently moved into new pastures.

Thus, some form of rotational grazing is desirable. The correct acreage per horse changes with the season as well as with other factors. However, a good rule of thumb in the humid region is to provide at least 1 acre of good-quality pasture per horse. Then set up five or six paddocks, letting the horses graze first in one area for about one week, and then change them to another. This system helps to keep the legumes and grasses growing better and increases the carrying capacity per acre. Furthermore, by rotating the grazing pattern you can break the life cycle of some parasites.

In regions of low rainfall more than 1 acre per horse will be needed if the pasture is to supply a significant amount of feed. The amount of land needed per horse varies widely with variation in rainfall, soil types and fertility, and pasture species.

Clipping your pastures regularly during the growing season is also an important management practice. Clipping at a height of 1½ to 2 inches after horses are moved to a new paddock helps to control weeds, prevents grasses from heading, and, in general, keeps the vegetation in a more palatable condition.

Fig. 15.2—Pastures for horses should be well-fenced in order to prevent injury.

Dragging pastures

Dragging pastures with a chain or link harrow at least once a year (in the fall) is also important. Dragging helps to spread animal manure. This helps to reduce the parasite population by exposing these parasites to air and sunlight. In addition, dragging helps smooth over areas dug up by horses' hoofs on wet soil.

Horses thrive on good pastures, their "natural habitat," but they still require proper management and care. Pastures should be well drained and safely fenced. Pits, stumps, poles, and places dangerous to horses should be eliminated or guarded. Pastures should be kept free of wire, trash, idle farm machinery, etc.

Ponies and other horses generally do not develop problems such as bloat and founder on pastures. Care should be taken to accustom horses to new and lush pastures in order to prevent scouring.

All animals should have continuous access to shade, fresh clean water, salt, and a simple mineral mixture. Animals should be checked daily. The veterinarian should be consulted for preventive medicine and parasite control recommendations. Local county agents and vocational agriculture teachers can be of assistance in recommending con-

Fig. 15.3—Neat fences improve both the function and beauty of good pastures.

trol measures for flies and external and internal parasites. The teeth and feet should be inspected at regular intervals. Hoof care—trimming, resetting shoes, etc.—is just as important in the field as in the stable.

It is best to graze various age groups separately. For example, yearlings should be grazed separately from older horses.

Certain practices in using pastures are extremely important in the control of internal parasites. The following will assist in reducing parasite infestation on the pasture and thus lower the number of parasites a horse will ingest while grazing. (1) Do not overstock, since parasite problems often occur on very short grass. (2) Rotate pastures as much as possible. (3) Since most of the common parasites are specific to a particular type of livestock, follow horses with cattle or sheep on the pasture. (4) Graze young animals (weanlings and yearlings) separately from older horses. (5) Do not spread fresh horse manure on pastures grazed by horses. (a) Store or compost horse manure away from horses for at least two weeks before using it. (b) It is best to spread horse manure in thin layers on fields that will not be used for grazing.

CHAPTER 16

Irrigated Pastures

Farmers irrigate pastures for several reasons:

1. To provide more herbage of better quality per acre.
2. To smooth seasonal variations in production (including emergency plantings).
3. To extend the growing season.
4. To maintain desired ratio of legume to grass stands and prolong stands in general.

The factor that determines whether an irrigation system is profitable is: Will the extra livestock products resulting from the additional grazing pay for the cost of irrigating? The cost of irrigating pastures and forage crops may be higher than for most other crops since they have a longer growing season. The value of pastures and forages relative to other crops is lower, so the farmer must get bigger yields (and higher quality) if irrigating them is to be profitable. Irrigation must be supplemented by use of high-yielding forage species, fertilization (usually), and soil management practices including provisions for drainage, erosion control, and land leveling in some instances. The irrigation system selected must fit the individual farmer's land and situation.

Rotational grazing of irrigated pastures is a requisite since the movement of animals over wet soil is harmful to the plants and tends to puddle the soil. That is, water should be applied after the animals are removed, thereby allowing the plants to recover more rapidly from adequate water supply.

Activities Which Involve Approved Practices

1. Supplying water.

Fig. 16.1—Heavy equipment is often needed in order to get the ground into as good a shape as possible prior to seeding flood-irrigated pastures. Laser-beam leveling is becoming increasingly popular in ground preparation when exact slope is required.

2. Obtaining necessary information.
3. Considering irrigation systems.
4. Determining costs and benefits.
5. Combining irrigation and fertilization.
6. Tailoring the system to the individual farm.

1. Supplying Water

The water for irrigating may be supplied as surface water from lakes; reservoirs; or large, deep ponds; or from streams or rivers. Another source might be ground water, such as deep wells or shallow wells (sand points) in areas having continuously recharged high water table conditions.

Check water supply

In any event, the person considering irrigation for a pasture should check with a representative of the State Water Survey Division and, in most instances, with the local Soil and Water Conservation Service representative to determine feasibility of use of the proposed water supply.

Few people realize the large amounts of water involved in irrigation of pastures. For instance, just to supply 1 inch of water to an acre, disregarding losses, requires 27,000 gallons. Due to content of salts, some water sources are simply not suitable for use in irrigation.

2. Obtaining Necessary Information

There is need to know the permissible rates of water application (based on infiltration and absorption capacity) to use on soils. On light sandy soils, such as sandy loams and sands, the application per hour may be from ½ to ¾ inch, while on heavy textured soils, such as sandy clays, silty clays, and clays, the rate of application should be from ⅒ to ¼ inch per hour.

As the slope of the land increases, particularly for sprinkler irrigation, there should be a reduction in the water application rate. On a slope of 9 to 12 percent the application rate would be only 60 percent of that for land having 0 to 5 percent slope and on land having over 20 percent slope, only 25 percent, comparatively.

The water-holding capacity of a light sandy soil may be only approximately an inch of water, while some heavy textured soils (clay) may be able to retain three times as much per foot of soil.

Irrigating efficiencies vary from 80 percent in a humid or cool climate to 65 percent in a desert climate.

Different plant species vary in their overall water requirements and the depth to which the soil should be wetted on the basis of rooting characteristics of the forage species.

Know the law

The person considering irrigation must have information on the principles of water law for his or her region. Two major principles of water law are in use in the United States. One is the common-law *doctrine of riparian rights* governing the usage of water by landowners whose holdings border a water body, such as a stream. The owner of land adjacent to a stream is entitled to use the full natural flow, undiminished in quantity and unchanged in quality, and the next downstream riparian owner of course has the same right. Thus there would tend to be a limit on the amount of water an upstream or lake border owner could use dependent upon the intended water uses of the downstream or adjacent neighbors, respectively. As applied to ground water, a "quasi-riparian" right is based on ownership of land overlying a water-bearing formation. A riparian right does not depend upon putting the water to use and thus is not lost by nonuse.

The second principle of water law, the *doctrine of prior appropriation,* is based on appropriation and actual use of water that is declared to belong to the public; the person who is first in time has the better right. The right is generally limited to water used beneficially and generally is forfeited if water is allowed to go unused for a specified period. The rule of reasonable use was modified by the California Doctrine of Correlative Rights, which requires one landowner's use to be reasonable, in consideration of the similar rights of others, but it must be correlated with the uses of others in times of shortage.

The water usage picture is clouded by the attempted distinction between water of "watercourses," surface or underground, and so-called "percolating" underground water. One rule of law may be

Courtesy, USDA

Fig. 16.2—To keep costs low, the irrigation system must be used to capacity.

applied to surface streams and another to percolating waters in a state. Also, different rules may be applied to water in "watercourses" and "diffused surface waters," such as flood waters not in water channels. All of these situations arise due to currently inadequate laws on water usage, but even so the tentative irrigator must check for the best possible information from the responsible agency in his or her state.

3. Considering Irrigation Systems

Irrigation water may be applied to pasture in one of several ways:

1. By flooding, thus wetting all of the land surface (border, basin, or contour ditch).
2. By using furrows or corrugations, thus wetting only part of the ground surface (the siphon tube and gated pipe are modifications).
3. By sprinkling, in which land is wetted much the same way as by rainfall (usually involves a moveable tube with perforations or sprinkler heads, or a revolving boom with sprinkler heads).
4. By subirrigation, in which the surface is wetted little, if any, but the subsoil is saturated.

The *border* method of flood irrigation is suitable for a wide range of soil textures, but on finer-textured soils the intake would be very slow. It is generally used for pasture, hay, and grain crops on lands having a slope up to 3 percent. In this system, ridges (borders) must be constructed along the strips to be flooded, dependent upon the lay of the land, and often land leveling is involved, along with larger heads of water for sod crops.

The *basin* method of flooding consists of running large streams of water into level (or nearly level) plots surrounded by dikes or levees. This system can be used with production of pasture, hay, or grain but is generally used on smoother land to avoid expenses of leveling.

In *contour ditch* irrigation, the field is flooded from comparatively shallow ditches running along the contour of the land. The water moves down slope between ditches, and no levees are used; but considerable time is involved in preparing the land for this system and in applying the water efficiently.

Flooding type irrigation is used mainly in the western and southwestern states.

Furrow irrigation is generally used for row crops, and furrows are

Courtesy, USDA

Fig. 16.3—Adequate water is necessary for irrigation.

Courtesy, USDA

Fig. 16.4—The irrigation system should be tailored to the individual farm.

generally run directly down the slope, but they can be run somewhat on the contour to better control erosion. With row crops a furrow is found between each two rows, but for pasture this spacing would depend upon the slope, infiltration, and lateral water passage capacity of the soil. The siphon tube and gated pipe may be the arrangements for conveying the water to the furrows from the main water supply.

Sprinkler irrigation is used on land too rough or too shallow to be leveled, on very porous soils, and in places where water is too scant for effective flooding. This system is adaptable to all of the major farm crops. Water distribution is affected by wind. Power requirements are usually higher for this system, and the water must be clean and free of debris. Water may be supplied by moveable perforated pipes (often aluminum) or from sprinkler heads on such pipes, on a revolving boom, or on a center pivot overhead pipe and sprinkler head unit. Positive control of application rates and amounts can be had by good engineering design, such that run-off erosion and deep seepage are avoided. Cost for power, pump, pipes, and related equipment, as well as labor involved, makes this system more expensive than those mentioned earlier. Due to its flexibility of use, this system has very wide adaptability geographically but is largely used in subhumid areas of the Midwest.

Subirrigation is the process of irrigation from below the ground surface. It is accomplished by creating and maintaining an artificial

Fig. 16.5—Concrete-lined ditches conserve irrigation water and reduce weed problems.

water table or by raising the natural water table to the place where the plant roots can utilize the moisture by capillarity. There must be a permeable layer for transmission of water and a lower barrier to rapid downward movement, such as a tight clay layer or high water table, and the soil must have a slight slope away from the supply ditch. This method is used on a limited acreage of fine, sandy loam and muck soils of Florida and along the Atlantic coast to New Jersey.

In any event, the design of an irrigation system requires the services of a person with engineering training, who might be located through a reputable irrigation equipment dealer, the agricultural engineering department of your college of agriculture, or technicians of the Soil and Water Conservation Service.

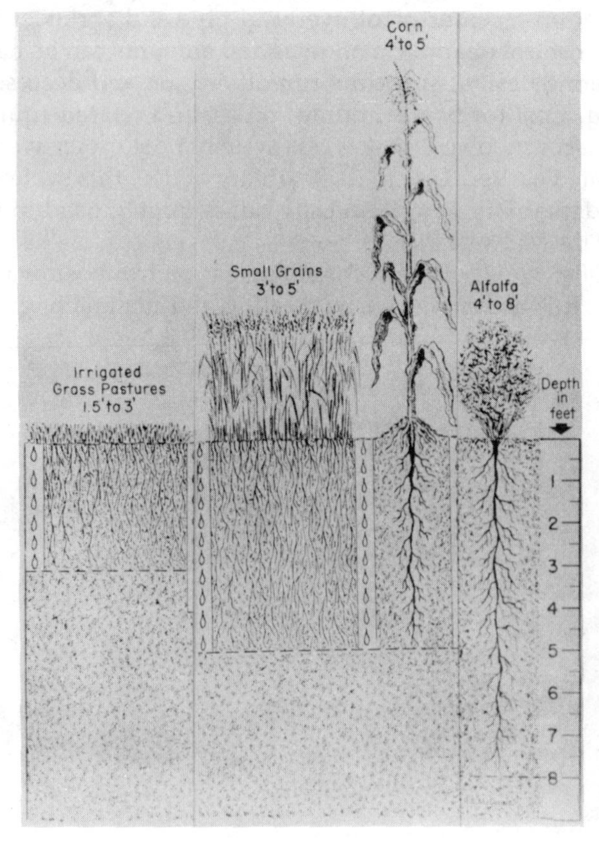

Courtesy, USDA

Fig. 16.6—Best results follow wetting the whole root zone.

Irrigation should supplement other good pasture management practices. Not only must other good practices be continued when irrigation is added, they may need to be improved in order to fully realize the production potential brought into the picture by incorporation of this particular management practice.

4. Determining Costs and Benefits

When the practice of irrigation is incorporated into a pasture management program, such action should result in more and better herbage, extend the growing season, aid in maintenance of a sustained yield of forage, and prolong the stands of forage species in that plot. Only if it contributes toward the foregoing can the practice be justified economically.

It is difficult to obtain statistics on the actual dollars and cents costs of irrigation applied specifically to forage production. Costs will vary with the irrigation method, fuel costs, and equipment used.

The Nebraska Agricultural Experiment Station found that when the acreage was large enough to make efficient use of equipment, the cost per acre with gated pipe or sprinkler irrigation was reduced greatly, although on very small acreages the costs of these two methods were much greater than the siphon tube system. On large acreages, the costs of water distribution for the three methods were found to be very similar.

Does it pay to irrigate forage crops? Michigan State University, using three 2-inch applications of water by sprinkler method on pasture, reported an 18 percent increase in forage, which just paid for the irrigation application. The increased forage production resulted in an increased milk production valued at just slightly more than the cost of applying irrigation water.

Reporting on supplemental irrigation of permanent pastures for beef cattle, the Virginia Agricultural Experiment Station indicated that the increase in carrying capacity of irrigated pastures was more important than total liveweight gain of individual steers. At this station, irrigation increased the productive life of the pasture because shallow-rooted species, such as Ladino and white clover, were more productive under the more favorable moisture conditions from irrigation.

Concerning irrigated pastures in the Willamette Valley and southern Oregon where heavy nitrogen applications were used by

Courtesy, Vemeer Mfg. Co.

Fig. 16.7—Pasture irrigation using a boom type sprinkler.

dairy producers, Oregon State University reported pasture growths equivalent to 11 tons of hay. It was stated that such yields were impossible without plenty of water and plenty of nitrogen.

At the Dixon Springs Experiment Station in Illinois, it was found that irrigated (supplemental by sprinkler) Ladino clover–grass pasture produced 1¼ tons per acre more dry matter than the non-irrigated pasture, resulting in an increased animal carrying capacity (largely beef) of 71 pecent. Irrigation was effective in maintaining legumes as 23 percent of the ground cover during the last year of a seven-year trial. Legumes in a legume-grass pasture produced more than 50 percent of the forage during the summer.

5. Combining Irrigation and Fertilization

Many fertilizers may be applied efficiently to almost any crop through irrigation water at any time during the growing season. Since the increased production from irrigation normally calls for increased application of plant nutrients, they may well be applied through the irrigation system where possible.

The water solubility and chemical nature of most nitrogen salts and solutions lend them to application through irrigation waters, unless they contain free ammonia, as in anhydrous ammonia. Potash fertilizers are water-soluble and therefore could be applied in the irrigation water, but this method would be used only when additions of potassium were needed since most of the needed potash requirement would have been broadcast before seeding. Phosphorous fertilizers should be applied before or at planting time because many of them are only partly water-soluble. Once applied, regardless of the form, the phosphate is rapidly converted to insoluble forms, so it doesn't move to any degree into the root zone.

The fertilizer should be dissolved in water in a container and then introduced into the pipeline in a controlled manner while irri-

Courtesy, Aluminum Company of America

Fig. 16.8—Gated aluminum piping for irrigation is easily moved from one position to the next.

Courtesy, USDA

Fig. 16.9—Light aluminum piping has been a boon to sprinkler irrigation.

gation water is applied. Aside from absorption of plant nutrient ions by the colloidal matter of the soil, the depth of fertilization can be regulated by the amount of irrigation water applied. Application of excess irrigation water could leach some nutrients beyond the root zone, but that is not very probable where there are good stands of forage species.

6. Tailoring the System to the Individual Farm

After determining whether an irrigation system is economically feasible for your farm or a particular enterprise in your farm operations, consult a reputable firm or the appropriate state agency to secure the services of a person with experience in irrigation engineering to design an irrigation system that will supply your needs.

The annual costs of water distribution by any irrigation system should be estimated along with costs for land grading (where necessary) and upkeep costs. Included should be interest on investment, depreciation charges, taxes, and repairs on all irrigation equipment, as well as human labor and tractor labor used in irrigating.

In addition, to utilize such a system effectively, you must:

1. Use an adapted mixture of grasses and legumes.
2. Apply the right kind and amount of fertilizer.

Table 16.1

Average Cost per Hour to Operate Agricultural Pumps[1,2]

Horsepower	KW Input	Service Charge		Average Cost per Hour at Various Annual Operating Hours[3]		
		Monthly	Annually	500	1,000	1,500
5	4	$ 5.50	$ 66.00	$ 0.48	$ 0.42	$ 0.40
10	8	8.50	102.00	0.90	0.80	0.77
15	13	11.50	138.00	1.39	1.25	1.20
20	17	14.50	174.00	1.80	1.63	1.57
25	21	17.50	210.00	2.21	2.00	1.93
30	25	20.50	246.00	2.63	2.38	2.30
40	33	22.30	267.60	3.36	3.09	3.00
50	41	27.10	325.20	4.15	3.83	3.72
60	50	32.50	390.00	5.05	4.66	4.53
75	62	39.70	476.40	6.25	5.77	5.62
100	82	51.70	620.40	8.25	7.63	7.42
125	103	64.30	771.60	10.34	9.57	9.32
150	123	76.30	915.60	12.34	11.43	11.12

[1]*Source:* University of California Extension Service.

[2]Electricity rates will change as time passes, so this table will only be relatively accurate. However, the reader will find the table useful in comparing costs relative to horsepower used.

[3]Average costs are based on PA-1 Rate effective Jan. 1, 1982, and include annual service charges.

3. Prevent or control undesirable weedy plants.
4. Manage grazing to utilize the additional forage produced.
5. Provide for a continuous supply of water to the plant species used.

Your county agricultural agent can be of great assistance in working with the preceding items.

CHAPTER 17

Emergency Pastures for Drought Conditions

Livestock farmers not having irrigation systems often follow the recommended practice of planting a limited acreage to crops that produce summer pasturage under drier-than-normal conditions. This is standard practice for many dairyfarmers who find that some or all of their permanent pastures drop in productivity in the hot, dry months.

Activities Which Involve Approved Practices

1. Growing plants on soils low in moisture.
2. Managing pastures before dry weather.
3. Utilizing deep-rooted plants.
4. Utilizing vinelike crops.

1. Growing Plants on Soils Low in Moisture

There are three kinds of plants well suited for grazing purposes during drought periods.

Sudangrass

Probably the most common pasture grass used for emergency pasture in humid areas is sudangrass and sorghum-sudan hybrids. Sudangrass has been discussed already in Chapter 4, and the reader is

Courtesy, USDA

Fig. 17.1—Sudangrass and sorghum-sudangrass hybrids provide abundant grazing during the hot summer months. Some poison danger exists in the fall after frost occurs.

referred to that chapter for the approved practices in establishing and maintaining sudangrass or sorghum-sudan hybrid pastures.

The sorghum-sudan hybrids can produce good quantities of quality livestock feed as demonstrated by the Michigan data presented in Table 17.1.

Bermudagrass

Another crop that grows well under hot, dry conditions is bermudagrass. It also has been discussed as one of the more important pasture grasses (see Chapter 4). With the introduction of the two high-yielding bermudagrasses, Coastal and Midland, the use of this grass as an insurance crop against hot, dry weather expanded markedly. Bermudagrass has the advantage over sudangrass of being a perennial, thus not requiring annual seeding. Chapter 4 lists the approved practices for establishing bermudagrass pasture.

Table 17.1

Animal Performance per Crop[1]

Crop, 3 Acres	Lbs. Dry Matter per Day				Lbs. Milk per Day	Lbs. Gain per Week
	Forage	Hay	Green	Total		
Sudax (sorghum-sudan hybrid)	17.2	2.3	10.5	30.0	32.8 (100%)	1.1
Piper sudangrass	19.4	2.1	10.0	31.5	31.9	3.0

[1]Performance of 10 Guernsey cows at the W. K. Kellogg farm, Battle Creek, Michigan, on a 9-acre field used for green chopping.

Grain sorghum

Grain sorghum has long been regarded as a crop that does well under low moisture conditions. This crop is used primarily for grain production, but when it is grazed in the dough stage it can provide a tremendous tonnage of pasture herbage. Because it's an annual, a new seedbed must be prepared each year. However, satisfactory stands are easy to obtain.

Seed treatment is recommended to control rots and to guarantee germination. This crop is planted in rows, and planting can be done with a corn planter equipped with proper planter plates. Otherwise, the crop can be planted with the grain drill. General fertilization practices recommended for corn are applicable to grain sorghum.

Grain sorghum is a versatile crop and one that can easily stand more attention in this country. If favorable weather for permanent pastures returns after the sorghum crop has been planted, the farmer can profit by ensiling the sorghum or harvesting it for grain.

2. Managing Pastures Before Dry Weather

For the most part, under range conditions, cattle and other livestock feed on grasses which make early spring growth. Because these areas are extremely dry, the forage loses little of its nutritive value. Even under humid conditions, some farmers grow a crop like soybeans or soybean-millet combinations for grazing during critical drought periods. Here, most growth is made before dry weather hits.

And, because of this early growth, there is considerable forage available for hot-weather grazing.

Sometimes orchardgrass and tall fescue are permitted to grow during the spring months without being grazed down. They both grow to considerable height and produce tremendous amounts of forage. Then if dry weather comes along, the cattle are turned into the field where growth has already been made during more favorable climatic conditions. Such forage is not as palatable as the shorter, more succulent growth but it is certainly better than no grass at all.

Fields set aside for what might be called "deferred grazing" can be used for hay or for seed production if weather conditions are favorable for pasture growth.

3. Utilizing Deep-Rooted Plants

The green, healthy appearance of alfalfa during a drought is a common sight. Alfalfa actually requires a tremendous amount of water, but because of its very deep root system, it is able to obtain water from a greater soil mass than most shallow-rooted plants. As a result, alfalfa stays green and succulent when shallow-rooted plants often wither or even die. Approved practices for alfalfa growing have already been covered in Chapter 4.

Sericea lespedeza

This lespedeza is a deep-rooted legume which withstands drought under southern conditions. It is not as palatable as alfalfa but has the advantage of growing on soils not so well stocked with lime or fertility. Animals when first turned onto sericea give evidence of not liking the forage. Once they become accustomed to it, however, production levels are satisfactory.

Overseed with clover

In the South, sericea fields are often overseeded with hard-seeded crimson clover. The crimson clover furnishes grazing during the fall, winter, and early spring, and sericea fills in during the hot, dry summer periods.

Like all other crops, a firm seedbed is desirable.

The seed should be inoculated

While sericea will make some growth on poor land, if a satisfactory forage growth is to be produced, an application of about 500 pounds per acre of a fertilizer such as 6-12-12 should be made at seeding time. Annual application of 500 pounds of 0-14-14, or similar fertilizer, per acre should be made in the early spring.

Sericea should be seeded broadcast on a well-prepared, firm seedbed in March or April at the rate of approximately 40 pounds of scarified seed per acre. The crop should not be grazed the first year, but if grasses or weeds interfere with growth, they should be clipped with a mower, the blades being raised as high as possible.

Grazing should start the spring of the second year when the shoots are 4 to 6 inches high.

Courtesy, USDA

Fig. 17.2—Sericea lespedeza is a good summer crop for the South.

Crimson clover fits in well following sericea

Many farmers in the south are finding it profitable to seed crimson clover in old stands of sericea. When this is done, crimson clover seed should be seeded immediately following the spring harvest or after the sericea is grazed down to about 2 or 3 inches. When crimson clover is top seeded on sericea, an application of approximately 500 pounds per acre of a 0-14-14, or similar fertilizer, should be applied at seeding time.

4. Utilizing Vinelike Crops

Kudzu is a deep-rooted, drought-resistant, annual legume that produces a coarse growth of leafy vine. The plants are killed back by the first frost, but, even so, kudzu can provide a tremendous amount of grazing during the summer months when many plants go dormant under drought conditions.

Kudzu has become a weedy plant in non-grazed areas. Extreme care should be taken to keep kudzu in the intended pastures.

Courtesy, USDA

Fig. 17.3—Kudzu should be grazed lightly until well established.

Prefertilize

The South Carolina Extension Service recommends that the land to be seeded to kudzu be prepared during the winter by breaking out deep furrows 25 feet apart and applying in these furrows generous amounts of phosphate and potash with lesser amounts of nitrogen.

The next step is to back-bed on each furrow until there is a low, flat bed 10 feet wide. When this bed has become firm, crowns are planted 3½ feet apart in the row. Seedlings and crowns should be planted in late winter or early spring while the plants are dormant. The soil should be firmly packed about the roots.

Kill weeds

Newly established kudzu plants require cultivation to keep down competing grasses and weeds until the space along the row is completely covered with vines.

Defer grazing

The field should not be grazed until the plants are well established, usually in the fall of the third or fourth year. Once established, kudzu may be grazed intermittently during the growing season and even after frost. Overgrazing will kill kudzu, though. After the plants are grazed down they should be allowed time to recover completely before being pastured the second time.

CHAPTER 18

Winter Pastures

Winter pastures are popular in areas that have mild winter temperatures. Many grasses will make considerable growth as long as the soil temperature is above 40° F. Especially is this true in the case of the small grains—rye, wheat, oats, and barley; and for that reason, these crops are used extensively for winter grazing.

Activities Which Involve Approved Practices

1. Increasing winter pasture acreage.
2. Selecting types of winter pastures.
3. Examining the economics of winter pastures.
4. Grazing winter pastures.

1. Increasing Winter Pasture Acreage

The acreage of land devoted to winter pastures in the South and in the Northwest is many times what it was just a decade ago. Using new practices and better plant species and varieties, brought about by agricultural research, it is now possible to have winter pastures where winters are mild—pastures nearly as productive as summer pastures.

2. Selecting Types of Winter Pastures

Winter pastures can be grouped into four different types, each with certain advantages and, of course, disadvantages. They are (1) temporary pastures—seeded annually on prepared seedbeds; (2) volunteer pastures—annual crops which come back from seed; (3) permanent pastures—perennials like tall fescue and orchardgrass that grow the year around in the warmer areas; and (4) field residues—usually surplus crops or forage. These different types of pastures will be discussed individually.

Temporary winter pastures

Temporary pastures include all fall sown grains, whether utilized in part or wholly for pastures. Ryegrass and crimson clover, alone or in mixtures, make good temporary pastures. Other legumes and grasses are often used. Oats is the small grain most commonly used because it is so widely adapted.

Many farmers prefer to seed a mixture of the small grains—rye, wheat, oats, and barley. They believe, with some justification, that this mixture provides more forage, especially on fields where variable soil conditions exist.

Ryegrass has the distinct advantage and value of forming good

Courtesy, USDA

Fig. 18.1—Small grains grow late in fall and early in spring when most pasture crops are dormant.

sod, which allows pasturing when soil-moisture conditions prevent grazing on small grains. It also provides grazing later in the spring than most small grain crops.

Numerous experiments have shown that approved practices lead to increased yields, early maturity, and improved nutritive value of temporary pasture plants.

These approved practices are:

1. Prepare the seedbed early.
2. Seed early.
3. Fertilize heavily as with 400 to 800 pounds of a 16-16-16 or other suitable fertilizer per acre at seeding time.
4. Apply lime if needed.
5. Cultipack to encourage early germination.
6. Top-dress at least 50 pounds of nitrogen (equivalent to 150 pounds of ammonium nitrate) shortly before starting to graze. Make another application before cold weather, and still a third in early spring.
7. Select the best-adapted species and varieties suited to local conditions.
8. Don't seed too thinly and not too heavily—greater returns for

Fig. 18.2—The land is prepared annually for temporary winter pastures.

money invested are obtained by good seedbed preparations and the use of sufficient fertilizer.

9. Regulate grazing to allow herbage accumulation in the fall months for use during mid-winter.

The value of good management practices is not fully appreciated. Proper fertilization, for example, will not only increase total winter grazing but will also stimulate growth in the fall so that 30 to 40 percent of the entire season's forage is produced in the fall and early winter periods. Without good fertilization, less than 15 percent of the total forage is produced before spring.

Increase nitrogen

Research findings indicate that greater forage production and higher profits result as the rate of nitrogen fertilizer is increased. Timely and adequate nitrogen top-dressings also encourage early growth and increase the nutritive value of the herbage by raising its protein content. In Georgia, where 16 pounds, 32 pounds, and 48 pounds of nitrogen per acre were applied as top-dressings monthly, the winter herbage contained an average of 16.7, 20.0, and 23.6 percent crude protein, respectively.

The chief advantages of temporary winter pastures are:

1. They provide rapid, early growth following early seeding and proper fertilization.
2. The forage produced is especially high in nutritive value.
3. In many cases, a grain crop can also be harvested from a field grazed earlier in the season.

The disadvantages are:

1. The cost is relatively high because of the need for annual seedings.
2. The establishment hazards encountered in seeding any crop recur annually.

Volunteer winter pastures

The plants adapted to volunteer winter pastures are mostly annuals that produce seed in late spring. The seed lies dormant during the summer, germinates in late summer or early fall, and produces herbage during winter and spring.

Crops of this type include hard-seeded strains of crimson clover, Caley or singletary peas, bur clover, button clover, subterranean clover, and to a small extent, cheat, rescuegrass, and ryegrass.

Rescuegrass volunteers from seed, but some plants tend to be perennials. To a lesser extent, the same is true in the case of ryegrass.

Follow cultivated crops with volunteer

Often, volunteer winter legumes are used in crop systems where they follow cultivated crops such as grain sorghum, cotton, or corn. The legumes germinate in the late fall, and grain or ryegrass may be sown early to increase the amount of early winter herbage produced. On some farms, the land is disced early in the fall before the volunteer crop comes up. This usually leads to earlier and higher production.

Fertilization generally includes the use of lime, phosphate, and potash for the legumes. Annual fall applications are preferable for best results.

The advantages of volunteer winter pastures are:

1. The cost is lower than the cost of temporary winter pastures, since annual seedbed preparation is not necessary.
2. On many farms, the cash seed crops obtained after grazing stops are worth as much as or more than the value of the grazing.

The disadvantages are:

1. The carrying capacity is lower than that of temporary winter pastures.
2. There is usually not an abundance of early forage—most of the production comes in late winter and early spring.

Permanent winter pastures

Tall fescue appears to fulfill the need for a widely adapted, productive winter growing perennial grass. This grass is more widely adapted than orchardgrass or bluegrass and provides more grazing in mid-winter. Once tall fescue is firmly established, it has a high carry-

Fig. 18.3—Large amounts of fertilizer, especially nitrogen, are required to keep winter pastures growing vigorously.

ing capacity. Two to four cows or more per acre can be carried for periods of three to six weeks.

In Georgia, yield measurements on one-year-old fescue showed that green forage available for grazing in late November amounted to 6,000 pounds (dry weight) per acre. By comparison, early sown and heavily fertilized temporary pastures yielded less than 1,000 pounds at the same time.

A system that was popular in the 1960's was to stockpile tall fescue. The May-June and June-July crops were harvested as hay in small round bales and were left in the field. The August growth of tall fescue was allowed to accumulate. This served as the winter pasture. The pasture was partitioned off to give the cattle about two weeks' supply at any one time. Beef cows were rather efficient in using the small round bales and the stockpiled fescue. This system has been replaced by the big (1,000-3,000 pound) round baler. The big round bale system requires feeding in a structure (ring, gate, or box) to prevent substantial waste. The big round baler can bale hay or pasture clippings faster than the small baler, and mechanized handling has been developed. Large round bales have a lower percent of spoilage than the small bales when both are stored outside.

Courtesy, New Holland Division of Sperry Rand

Fig. 18.4—Cultivated crop residues can extend the grazing season.

The big round bales are usually moved off the pasture or to the edge. The bales are fenced off to keep cattle from them. The bales are usually fed during the winter or during drought periods. The feeding may occur in dry lot, on other crop residues, or in the pasture from which the bales were harvested.

Many producers harvest a seed crop from the fescue and then stockpile the regrowth for winter use. If a good seed crop is desired the following year, close clipping of the fescue soon after seed harvest is essential. If 40 to 60 pounds of nitrogen is applied after seed harvest and the regrowth is utilized, then an additional 40 to 50 pounds of nitrogen applied in December or January will be necessary for a good seed crop the following summer.

Next to tall fescue, orchardgrass is perhaps the best adapted perennial grass for winter grazing conditions. It does not, however, produce the tonnage of herbage that tall fescue does. Approved practices for maintaining productive permanent winter pastures include:

1. Fertilize properly and use lime where needed. The use of 400 to 600 pounds of fertilizer, as a 12-12-12 or a 16-16-16 is recommended at planting time.

2. Top-dress with at least 50 pounds of nitrogen per acre in the fall, spring, and mid-summer of the first year.

3. Seed a legume, preferably alfalfa or red clover, with the grass to provide a source of nitrogen and improve the palatability and nutritive value of the herbage.

4. Seed in the fall. Do not overgraze the first year or anticipate year-round grazing. Most perennial grasses may be slow to establish and most will not furnish much grazing in dry, hot months.

Field residues as winter pastures

This type of winter grazing is usually not green grazing. It does, however, furnish feed when most needed at low cost. Where this type of grazing is practiced, farmers permit the crop to grow un-

Fig. 18.5—In the warmer regions, perennial grasses can provide grazing most of the year.

disturbed during the more favorable months. After growth is stopped because of low temperatures, the animals are turned in to harvest the forage already produced. There is some loss of feed value from leaching, but this system of grazing can be used profitably by many beef cattle farmers and by dairy farmers supporting dry cows. The grasses commonly used include fescue and orchardgrass. Legumes may or may not be included in the mixture.

3. Examining the Economics of Winter Pastures

The expected profits from winter grazing, of course, vary with the livestock enterprise and the climatic conditions of the area. Naturally, more production is obtained in years with mild winters than during cold years.

4. Grazing Winter Pastures

Many livestock farmers, even those who practice winter grazing, occasionally "run out" of pasture in mid-winter. The Georgia Agricultural Experiment Station studied the growth rate of forage in winter as influenced by management and fertilization practices. Fig. 18.6 shows the seasonal dry matter production of a temporary pasture mixture seeded early and fertilized with 48 pounds nitrogen, 64 pounds phosphate, and 48 pounds potash.

The two horizontal lines show the dairy forage requirements needed to provide the total digestible nutrients for maintenance of a 1,000-pound animal and production of 20 pounds of milk daily. It is seen that little forage growth was obtained from mid-December to mid-February under conditions which would approximate continuous grazing with minimum nitrogen top-dressings. Much more growth took place during this period when less frequent defoliation was practiced and more nitrogen was used. Even then, there was not sufficient growth until late February to feed one cow per acre sufficient nutrients for normal milk production.

The forage used in this experiment provided more than enough protein per acre throughout the winter to supply the needs of one cow, giving 20 pounds of milk. Winter grazing crops are usually higher in protein than warm-weather crops.

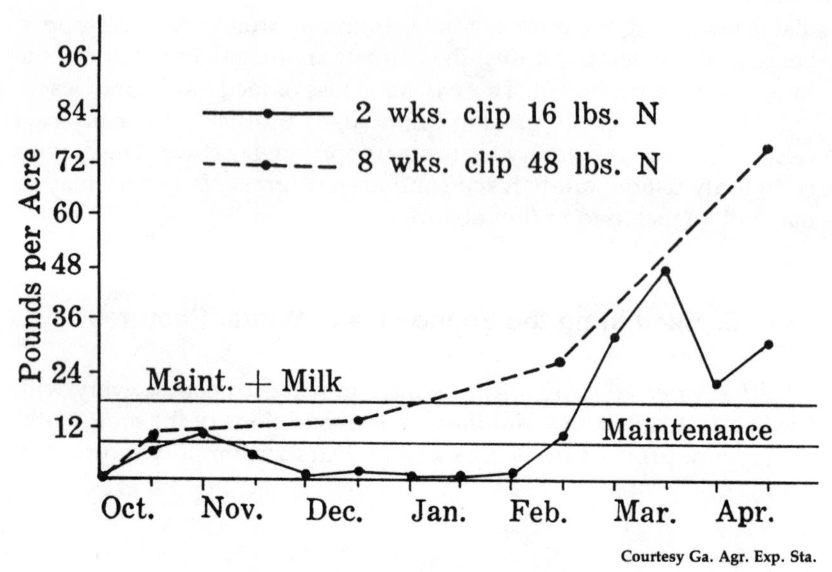

Courtesy Ga. Agr. Exp. Sta.

Fig. 18.6—Daily dry forage produced from winter pastures.

Tall fescue and other winter grazing crops usually contain less protein than small grains. Nevertheless, practically all winter grazing plants with the exception of the residue type are lower in fiber-and-energy-supplying nutrients and higher in protein than most warm-weather pasture crops.

CHAPTER 19

Mechanical Harvest of Forages

Regardless of how hard the farmer tries, there will be periods when herbage production is more than the animals on the farm can consume, and there will be other periods when pasture growth does not keep up with the animals' needs. This happens for two reasons. First, variations in weather, especially rainfall, greatly influence the rate of growth of most pasture grasses and legumes. Secondly, most grasses and legumes are cool-season crops and naturally grow faster in spring than in mid-summer. There is also another flush growth period in the fall, when weather cools and moisture supply increases from fall rains.

If farmers reduce their pasture acreage to a point where the animals keep up with the flush growth periods, they are faced with a deficit during the slow growing periods. What they try to do is to grow a little more than is needed during the good parts of the season, then make provisions to harvest the surplus to tide the animals over during the distress periods.

The making of hay from surplus forage has been practiced for many years.

Nevertheless, the quality of the hay often is quite low. In addition to being used to make hay, the forage may be conserved for later feeding by storing it in a silo. This chapter deals with the approved practices associated with making both quality hay and nutritious grass silage.

Activities Which Involve Approved Practices

1. Harvesting to obtain maximum feed value.

2. Making and storing forage economically.
3. Making and storing grass silage.
4. Wilting to produce grass silage.
5. Using preservatives in making grass silage.
6. Deciding which silo for a farm unit.
7. Considerations in making grass silage.

1. Harvesting to Obtain Maximum Feed Value

Proper timing in hay-making is very important to minimize field losses. Too much hay is raked and baled under conditions that lead to extensive losses of quality and yield. Animals receiving lower-quality hay will require more hay for body maintenance and production of meat, milk, wool, or work. Cubing and pelleting of hay can use various qualities of hay. Most cubes and pellets are marketed on a minimum crude protein content.

Agricultural engineers at the University of California conducted research on hay-making. The harvest was handled in four different manners and the hay quality was measured by animal gains. The treatments were:

(1) Handled properly: This treatment was raked at 5 mph when the hay was tough with dew and had a moisture content of 40 to 50 percent. It was baled or wafered in the morning after free moisture from dew had disappeared from the tops of the windrows but before the hay was dry enough to cause appreciable leaf shatter (13 to 16 percent moisture content).

(2) Raked dry: This treatment was raked at 5 mph during the afternoon at moisture contents conducive to serious leaf shatter. This dry condition first occurred 1½ days after mowing for conditioned hay (12 to 15 percent moisture), and at 2½ days for unconditioned hay (10 to 12 percent moisture). The hay was baled properly as in (1).

(3) Packaged dry (baled or wafered): This treatment was raked properly as in (1), but was baled or wafered during the afternoon at 5 to 7 percent moisture content (following an early-morning moisture of about 15 percent).

(4) Raked and packaged dry (baled or wafered): This treatment was raked as in (2) and baled or wafered as in (3).

The hay was second-cutting irrigated alfalfa. All hay was mowed during a four-hour period beginning at 8:00 a.m. A smooth roll-type

Courtesy, Deere & Co.

Fig. 19.1—Special equipment is needed to bale and handle the large round bales of hay. This method saves time, labor, and harvesting costs. Bales may weigh as much as 1½ tons.

crusher was used on the conditioned hay 20 minutes after mowing. Raking was done with an oblique reed-head rake at 5 mph. Four passes were made putting the hay into one large windrow for baling and two small windrows for wafering. Baling was done with a two-wire automatic baler set to produce 16 × 18 × 42-inch bales weighing about 90 pounds each. The waferer was a roll-type machine producing wafers 3 × 3 inches and about 1 inch thick, averaging 10 to 14 pounds per cubic foot bulk specific gravity. The baled hay was chopped, and part of the chopped hay was pelleted. A flat bed-type pellet mill with a ⅜-inch die was used to produce the pelleted alfalfa hay for the feeding trial. The hay was then fed to sheep, and the data on yield, quality, and feeding responses taken are shown in Table 19.1 and Table 19.2.

The California scientists came to these conclusions:

> Packaging hay at low moisture content and during the dry part of the day caused only 4 percent loss in yield, compared to a 25 percent loss when the hay was raked too dry. Hay that was

both raked and packaged dry yielded 35 percent less than hay that was handled properly. In each case, the loss of protein was somewhat greater than the yield loss, indicating that the reduction in yield was predominantly leaves. When the hay was fed

Table 19.1

Influence of Field Treatments on Yield, Hay Quality, and Feeding Response[1]

	Handled Properly	Packaged Dry	Raked Dry[2]	Raked and Packaged Dry[2]
Harvested yield per acre				
Dry matter, lbs.	2,920	2,800	2,170[3]	1,900[3]
Crude protein, lbs.	654	608	442	368
TDN, lbs.[4]	1,700	1,630	1,220	1,030
Lamb, lbs.	223	218	152	115
Net energy, Mcal[5]	1,590	1,550	1,150	970
Composition (dry-matter basis)				
Crude fiber (modified), percent	26.7	26.8	28.7	30.6
Crude protein, percent	22.4	21.8	20.4	19.4
Estimated TDN, percent[6]	58.3	58.1	56.4	54.6
Estimated net energy, Mcal per 100 lbs.[7]	54.5	55.4	52.9	51.1
Sheep data				
Number of sheep	18	18	18	18
Initial weight, lbs.	77	78	75	78
Daily gain, lbs.	0.19	0.19	0.15	0.15
Daily feed intake, lbs. dry matter	2.49	2.48	2.31	2.45
Adjusted daily gain, lbs.[8]	0.18	0.19	0.17	0.15
Gain per 100 lbs. feed intake, lbs.	7.5	7.8	6.7	6.1
Carcass data				
Dressing, percent	51.3	50.6	50.5	48.3
Carcass fat, percent	22.2	22.5	19.5	20.5
Grade				
No. choice	1	4	—	1
No. good	16	13	11	12
No. utility	1	1	7	5

[1]Results for conditioned and unconditioned, baled and wafered, and chopped and pelleted hay are combined for each field treatment in this table.

[2]Statistically significant differences from the hay handled properly (5 percent level).

[3]Treatment yield is significantly different at 1 percent level from all listed yields.

[4]Estimated as suggested by Meyer, J. H., and G. P. Lofgreen, *J. Animal Science*, 18:1233, 1959.

[5]Calculated from equations given by Garrett et al., *J. Animal Science*, 18:528, 1959.

[6]Estimated from crude fiber content.

[7]Estimated from energy gain, maintenance requirements, and feed intake.

[8]Adjusted to equal feed consumption in the analysis of covariance.

Table 19.2

Influence of Hay-making Practices on Feeding Response with Sheep[1]

	Handled Properly	Packaged Dry	Raked Dry	Raked and Packaged Dry	Average
Pellets (3/8-in. dia.)					
Average gain, lbs.	28.0	24.2	25.3	22.3	25.0
Feed intake, lbs. d.m.	314	290	282	291	294
Gain per 100 lbs. feed, lbs.	8.9	8.3	8.9	7.7	8.5
Wafers					
Average gain, lbs.	16.3	17.5	11.7	9.5	13.7
Feed intake, lbs. d.m.	239	249	220	228	234
Gain per 100 lbs. feed, lbs.	6.8	7.0	5.3	4.2	5.9
Chopped					
Average gain, lbs.	14.7	19.5	11.7	15.3	15.3
Feed intake, lbs. d.m.	233	245	228	253	240
Gain per 100 lbs. feed, lbs.	6.3	8.0	5.1	6.0	6.4
Average, all					
Average gain, lbs.	19.7	20.4	16.2	15.7	18.0
Feed intake, lbs. d.m.	262	261	243	257	256
Gain per 100 lbs. feed, lbs.	7.5	7.8	6.7	6.1	7.0

[1]The conditioned hay and unconditioned hay has been combined.

to sheep, the reduction in hay quality resulting from field treatments was similarly reflected in smaller weight gains. Wafered hay was equal to chopped hay as a feed; the same hay, pelleted, resulted in increased gains with sheep. Conditioning had little effect except in increasing the curing rate.

2. Making and Storing Forage Economically

There have been many improvements in forage-handling equipment. The shortage of reliable help and the high cost of labor have triggered the development of labor-saving forage- and hay-handling equipment. As a result, forage-making does not need to be a high-cost labor job on the farm.

Mower-conditioners and self-propelled windrowers made their appearance in the 1950's and 1960's. These machines were the forerunners of other labor-saving hay-handling equipment, such as bale ejectors, self-propelled and pull-type bale-loading wagons, and, more recently, giant balers, loose hay-stacking wagons, hay cubers, and hay-pelleting machines.

Baled hay

The baler has been with us for a long time, and it will no doubt continue to be a favorite way of making hay on many farms. Self-propelled bale wagons that pick the bales up from the field and stack them without the use of hand labor appeal to many farm managers. Big self-propelled wagons can haul bales up to a mile and stack as many as 3,000 to 4,000 bales in an eight-hour day. This is all done without any hand labor. The stacks are straight and neat—equal to the best job done by hand labor.

A recent development is the large round baler. The big round bales weigh from 1,000 to 3,000 pounds each. These balers pick up forage from the windrows until the desired bale size is attained. Most of these machines stop and eject the bale to the ground. It is possible to use the baler to haul the bales to the desired storage area, but economy usually favors keeping the baler working and picking up the bales with other equipment. The big baler can bale up to 15 tons per hour.

If the bales are to be removed from the field, bale-moving equipment that works on a three-point rear tractor hitch can handle the job efficiently. Such equipment has long arms that slide under the bale with a hook that grabs the top of the bale. One bale is handled at a time.

Some livestock producers leave the bales in the field and turn the animals into the field and let them eat the forage at the bale site. Limiting the number of bales available will increase the efficiency of utilization. A feeding ring or similar device will further reduce wastage by the animals.

Drying chopped hay with a hay finisher is an inexpensive system of harvesting hay for home use, and one that lends itself to mechanization. The handling system is well mechanized to the extent of

Courtesy, New Holland Division of Sperry Rand

Fig. 19.2—A modern self-loading and -unloading bale wagon.

getting the hay onto the dryer. Feeding is still a hand operation, the extent of which depends on the storage unit and the location of the feed bunks with respect to the dryer. Chopped hay storage units adaptable to mechanized feeding or self-feeding have been devised but have not been widely accepted. Also, the wagon-drying system which enjoyed some popularity in the eastern part of the United States in the 1960's was not accepted in the West. In areas with climates similar to California's artificial drying has been utilized mainly as insurance against spontaneous heating, whereas the East needs such drying to save its hay from rain damage. Wagon drying permits curing the hay in the wagon and then storing it in units adaptable to mechanized feeding. Artificial drying provides a means of maintaining optimum hay quality.

Loose-hay stackers

Loose-hay stacking machines have been widely accepted in the Great Plains states, especially in beef cow herd areas. The stacking

Courtesy, Hesston Corporation

Fig. 19.3—A loose-hay stacker.

frame is on wheels and pulled behind a tractor. A flail type blower propels the forage from the windrow to the stacking wagon. When the stack is built within the wagon frame, it is either hauled to the desired storage area or dropped on the spot. There is a stack retriever that picks up the stacks and hauls them to the feed lot or to any other desired location.

Loose-hay stackers are manufactured in several sizes. The smallest makes a stack that is about 7 feet wide, 8 feet long, and 8 feet high. Some stackers employ a hydraulic cylinder mechanism that packs the hay in the stack. Other stackers grind or chop the forage slightly to aid in making a tightly packed stack.

Cubed or wafered hay

Many hay growers, especially commercial growers, believe that hay cubes are the ultimate choice for complete mechanization of harvesting and handling the forage for feed. Almost every operation can be completely automated, which significantly reduces labor costs.

Cubes or wafers lower transportation and storage costs and may increase palatability, if low-quality hay is used.

Present cubing accounts for some 20 percent of the alfalfa hay produced in Arizona and New Mexico and 8 percent of the alfalfa harvested in California, Washington, and Oregon.

The self-propelled hay cubers are the most popular machines in the West. They were first introduced in 1967. Stationary cubers are used in the humid areas of the U.S. and have several advantages. These advantages include improved efficiency, longer operation, and greater flexibility. Furthermore, the stationary cubers can be used to cube stacked or baled hay in the off season. The stationary cuber is also more suitable for adding supplements to the cubes.

In a research study in Kern County, California, involving 160 calves, animals fed alfalfa cubes ate on the average 2.4 pounds more per head per day than did animals fed baled hay. The cube-fed animals gained 0.62 pound more and consumed 4.02 pounds less feed per pound of gain.

To summarize, the advantages of cubes include:

1. Lower transportation and storage costs.
2. Reduced labor in feeding made possible through automation.
3. Decreased nutrient loss, as animals are less selective in eating the forage.

Courtesy, New Holland Division of Sperry Rand

Fig. 19.4—Surplus herbage can be made into haylage or silage.

4. More palatability than meals or hay.
5. Less difficulty in getting animals on feed quickly.
6. Elimination of most of the dust.
7. Reduced losses from wind.
8. Higher utilization of nutrients by animals.

Pelleted hay

Pelleting hay is used to mechanize handling, reduce transportation costs, and blend hays of differing qualities. The hay must be ground before pelleting; thus the total cost of processing becomes appreciable. Feeding trials of good-quality hay made into pellets have shown that part of the cost of pelleting can be regained through increased animal performance. The balance must be paid by reduced storage and handling costs and by less waste of feed. Certainly, pelleted hay is the ideal form for handling roughage in transport, in storage, and for feeding because of its density and free-flowing characteristics. Improvements in mechanization for storage and feeding and increased knowledge regarding animal response to pelleted hay may result in a greater use of this method of hay handling.

3. Making and Storing Grass Silage

After many years of study, including detailed examination and analysis of hundreds of samples of grass silage, the Massachusetts Agricultural Experiment Station has concluded that good grass silage has the following characteristics:

1. A pH of 4.5 or less (the nearer 4.0, the better).
2. Low volatile base content (0.5 percent or less expressed as ammonia).
3. Lactic acid content of 3 to 5 percent or more.
4. Butyric acid content of 2 percent or less.

All these characteristics except pH are expressed on the dry basis, pH being the same either wet or dry.

Silage meeting these specifications is relished by the animals and has a high feeding value and pleasing odor. Silage not conforming to these standards usually possesses offensive odors and pH values in excess of 4.5.

Watch the moisture content

The moisture content of grass-legume silage is a very important factor that influences the fermentation process and finally the quality of preservation. The desired moisture content varies with the type of silo. The most simple silos are stack, bunker, and trench. The forage in these silos is exposed to more air and is not packed as densely as in tower silos. The moisture of grass-legume forage to be ensiled in these simple structures should be 70-75 percent.

Tower silos are of two basic types: unsealed and gastight. Most unsealed tower silos are of a concrete stave construction and are open at the top (although many have a roof to protect the stored forage from rain, snow, and sunlight). The walls and doors must be airtight when filled with forage if good-quality silage is made. Some 4-6 inches of spoilage at the top is the accepted cost for this type of silo. Most gastight silos are constructed of steel which is lined with a non-corrosive material. There are also some concrete silos that are considered gastight.

Courtesy, New Holland Division of Sperry Rand

Fig. 19.5—The "green chop" method of handling forage can be tied in with silage methods.

The moisture content of forage to be stored in the unsealed tower silo should be 60-70 percent. Forages may be stored at 50-60 percent moisture in the gastight tower silos.

Attaining a satisfactory moisture level in herbage cut from lush-growing pastures poses quite a problem. Almost without exception, the moisture content is in excess of 70 percent and often runs as high as 80 percent, especially when the mixture contains a large proportion of succulent legumes. There are two ways, however, to handle the problem. They are:

1. Using the wilting method—letting the excess moisture evaporate from the herbage after cutting and before ensiling.
2. Using preservatives or conditioners.

4. Wilting to Produce Grass Silage

Numerous farmers have produced excellent-quality grass silage by using the wilting method. Two to four hours in the swath, on a bright day, is usually sufficient to reduce the moisture content of the forage to 70 percent or less. On cloudy or humid days, of course, more time is required.

Because of the following conditions, the wilting method is impractical on many farms:

1. Weather conditions may not be suitable for wilting when the crop needs to be cut.
2. Taking time out for wilting poses a labor problem. Many farmers are reluctant to wait for a good wilting day once they have assembled their crews and set up the silo-filling equipment.
3. Wilting does not lend itself to the use of the modern field chopper.
4. Prolonged unsuitable weather conditions cause loss of food nutrients, may cause complete spoilage of the herbage, and retard the recovery growth of the pasture or hay crop.

Haylage

A distinction should be made between grass silage normally stored unwilted and with a moisture percentage of around 80, and haylage (low-moisture silage) where the plants are wilted, after being cut, to a moisture percentage of 50 to 60. Haylage can be stored only if air is completely excluded, so only sealed gastight silos are used for the preservation of forage near 50 percent moisture. Proper handling of low-moisture silage, or haylage, eliminates odors, requries no preservatives, provides abundant protein and vitamin D in the feed, and has few freezing troubles during the winter feeding period.

5. Using Preservatives in Making Grass Silage

Because wilting requires some time for drying in the field, some farmers prefer to use preservatives with fresh cut forage. Preservatives may be divided into two general groups. In the first group are those that add readily fermentable carbohydrates to the silage, thus encouraging lactic acid formation. The carbohydrates are usually starch or some form of sugar. Corn meal, dry whey, and molasses are typical examples.

The second group of preservatives are mild antiseptics. Their action reduces undesirable fermentation by quickly lowering the pH of the green crop. They are usually acids or acid-forming compounds. Included in this group are phosphoric acid, sodium bisulfite, liquid sulfur dioxide, and formic acid.

Conditioners

There is another group of what might be called conditioners. These materials absorb excess water from the green crop, thereby reducing the moisture content of the total mass. Examples of such conditioners are chopped dry hay, dried citrus pulp, and dried beet pulp. Ground cereal grains, including corn-and-cob meal, in addition to their preservative function also act as conditioners.

Molasses

For years, molasses was the principal preservative used in making grass silage. Although the molasses added carbohydrate it also added extra water. Since excess water is the principal factor to be avoided in making quality grass silage, the use of molasses has become less popular. In addition, there was always the problem of getting the molasses to "run" when the weather was cool.

Ground cereal grain

Cereal grain furnishes readily fermentable starch and also lowers the moisture content of the mass. Suitable grains are ground corn, hominy meal, ground barley, wheat, and oats. The amount to add, of course, depends somewhat on the content of the green crop. The right amount is usually about 150 pounds per ton. Assuming that the water content of the crop is 75 percent, the adding of 150 pounds of 12-percent-moisture grain reduces the water content of the total mass to 70.6 percent. This reduction in water content plus the boost given to lactic acid fermentation helps insure excellent-quality silage with a low pH and a pleasant odor. Such silage is relished by the animals.

Another advantage to ground grain is that it is easy to mix with the green crop. The quantity needed may be spread on top of the load of chopped herbage, or it may be metered with a device such as a fertilizer hopper geared to the blower.

There are those who insist that grains are too expensive to use as preservatives. It should be kept in mind, however, that most of the feeding value of the grains is preserved—they actually increase the feeding value of the silage. Research results indicate that 80 or 90 percent of the feeding value of the grain is preserved. Therefore, the cost of grain as a preservative is only a small fraction of the actual cost of the silage.

Phosphoric acid

Phosphoric acid is no longer extensively used for two reasons. First, its corrosive action on equipment and silo walls is a definite drawback. Secondly, its use produces silage of rather poor palatability.

Sulfur dioxide

Sulfur dioxide has marked antiseptic properties. It is marketed as a liquid under pressure in metal cylinders containing 150 pounds of liquid. It is applied to the crop in the silo with specially designed equipment, usually at the rate of 5 pounds per ton of green crop. An even application of the gas (which is difficult to obtain) is necessary for uniform silage quality. The application of the gas is an irksome task, especially if some of it escapes. Likewise, palatability of the silage, although reasonably good, usually does not compare well with ground grain or citrus meal.

With the newer and more easily handled sodium bisulfite, it is likely that liquid sulfur dioxide will become less popular.

Sodium bisulfite

Sodium bisulfite has basically the same action as liquid sulfur dioxide. It is, however, a crystalline powder which can easily be applied without special equipment. Its preservative action is due to the slow release of sulfur dioxide which takes place when the powder comes in contact with green herbage. Eight to 10 pounds of sodium bisulfite per ton of crop usually insures good preservation.

Formic acid

Formic acid has been used more in Europe than the United States. It is a good preservative but is corrosive to handling equipment and silos.

Distribute uniformly

There are metering devices on the market that can distribute preservatives or conditioners uniformly. However, a uniform distribution of granular forms of preservatives or conditioners can be obtained by sprinkling half the required amount on the green chopped forage when the wagon is half loaded and by scattering the other half on top of the load. When this is done and the wagon is unloaded by allowing the chopped grass to fall more or less vertically into the conveyor, the operation being speeded up by mechanical unloaders or by clawing the grass down with a rake or hay fork, satisfactory mixing is accomplished.

Table 19.3

Grass Silage Quality Studies at the Massachusetts Agricultural Experiment Station

Number of Lots	Treatment	Water in the Fresh Silage (%)	pH	Protein (%)	Fiber (%)	Volatile Bases (%)	Lactic Acid (%)	Butyric Acid (%)	Net Energy (Therms Per 100 lbs.)	Weight Losses (%)	Cost of Treatment per Ton of Silage Fed Out	Odor	Palatability
7	None	73.0	4.9	11.1	38.4	1.1	none	5.1	37.8	no data	—	Variable	Variable
30	Various preservatives	71.3	4.1	12.6	34.8	0.6	3.8	2.7	38.0	19.4	—	Variable	Variable
7	None	73.0	4.9	11.1	38.4	1.1	none	5.1	37.8	no data	—	Variable	Variable
20	Molasses	70.3	4.3	13.6	34.8	0.7	4.3	2.0	39.4	22.3	$1.25	Variable	Generally good
7	None	73.0	4.9	11.1	38.4	1.1	none	5.1	37.8	no data	—	Variable	Variable
7	Ground cereal grain	68.5	4.5	11.8	34.6	0.5	6.3	2.0	46.7	no data	—	Generally good	Excellent
1	None	71.8	4.8	13.6	36.7	1.1	none	4.1	39.0	39.8	—	Terrible	Poor
1	Hominy meal	69.3	4.2	11.5	30.1	0.5	10.1	0.3	46.6	14.4	$1.29	Pleasingly mild	Excellent
2	None	71.8	4.8	13.6	36.7	1.1	none	4.1	39.0	39.8	—	Terrible	Poor
2	Sulfur dioxide	75.3	3.8	13.8	31.4	0.3	5.0	3.5	44.5	21.1	$1.38	Peculiar	Fair
3	None (trench silo)	69.2	4.0	11.5	36.6	0.4	4.6	0.6	38.0	10.5	—	Variable	Fair to good
4	Sodium bisulfite	73.8	4.2	11.7	34.4	0.3	6.9	0.8	42.7	13.1	$1.07	Very mild	Excellent
5	None	73.0	4.8	11.0	37.3	1.2	none	5.5	40.0[1]	23.7[1]	—	Poor to fair	Fair
3	Citrus meal	72.3	4.4	12.8	34.0	0.5	1.6	2.1	41.5	19.3	$1.08	Variable	Good

[1]Some influence of the Harvestore silo here.

Citrus meal, citrus pulp, and beet pulp

These materials are conditioners rather than preservatives. They are low in moisture, carrying only 7 to 8 percent water as compared to 12 percent or more in ground grain. Their action is primarily one of lowering the moisture content of the total mass to a safe ensiling point.

6. Deciding Which Silo for a Farm Unit

Any structure used in the preservation and storage of forage crops may be called a silo. Green crops are packed in such structures at moisture levels high enough to convert the material to silage by fermentation. Silo types vary as to structural cost, life, labor requirements, and annual expense per ton of silage.

Upright or tower silo

An upright, tower, or vertical silo is usually cylindrical and can be constructed of concrete staves, formed concrete, double-walled glazed tile blocks, wood, sheet metal, brick, glass-lined steel plates, or galvanized sheet metal. A unit of this type may last for 25 to 50 years and have a fairly low annual cost per ton of silage. Ensiling losses are usually under 15 percent of the dry matter, and with an airtight or glasslined structure, this loss may be held as low as 5 percent.

Horizontal or trench silo

Supporting sidewalls or earth walls may be used, and the floor may be the soil or other material in a comparatively narrow trench. Cost of construction may be quite low for an unlined excavation, or a little more for a concrete-lined silo. The respective life spans may range from 5 to 25 years. A well drained site close to the feeding area is most desirable. Ensiling losses often run as high as 20 percent in these silos, making the annual storage cost per ton moderately high. Large silos which stack the silage 20 feet high or more have less loss of feed nutrients than shallower and smaller silos.

Fig. 19.6—Completely mechanized method of filling an upright silo.

Courtesy, USDA

Fig. 19.7—A trench silo provides economical storage for silage.

Half-in and half-out of the ground silo

Soil from the excavation is placed at either side of the linear excavation to increase the capacity and reduce costs of digging. There may or may not be retaining walls and flooring, and usually some covering material over the silage is required to lessen ensiling losses.

Stack silo

Forage is stacked directly on the ground or a simple base, such as a concrete slab. In this type of silo, surface spoilage provides a sealing blanket eventually. The initial cost per ton is very low, but the ensiling losses may run up to a third of the dry matter, making the per-ton cost quite expensive.

Above-the-ground or bunker silo

A silo of this kind may be built with or without semi-permanent or permanent walls and floors, but the bunker walls should be made of concrete or tongue and groove dimension lumber in order to provide an airtight seal. Adequate drainage is a requirement. Sidewalls should be sloped 1:8 in order to obtain proper packing. Life expectation is around 20 years. Ensiling losses are similar to those of trench silo while annual costs are comparatively high. Since livestock can feed directly from an above-the-ground or bunker silo, little or no unloading expense is involved.

Temporary silo

Here the retaining walls may be constructed of snow fencing and plastic films, woven wire fencing lined with weather-resistant paper, baled hay, or other material with some supporting and reinforcing vertical members. Such a structure lasts one to three years, although some of the structural materials may be used over again. Dry matter losses may run 20 percent. Plastic films (polyethylene) and bags are very effective in preserving silage but have a relatively high annual cost since the life span is so short. Mechanical loaders of plastic bag silos have made this option very practical for short-term storage plans.

7. Considerations in Making Grass Silage

Stage of maturity

Immature crops are higher in nutritive value than mature crops. However, yield, moisture content of the forage, and rate of recovery of the plants after harvesting must be considered when selecting the desirable stage for harvesting. Alfalfa, the clovers, and many major crops should be harvested for silage in the early bud to one-tenth bloom stage of growth.

Filling period

The filling period should be as short as practical. When the filling

Courtesy, A. O. Smith Co. (Harvestore Div.)

Fig. 19.8—For haylage or silage, good storage is an important asset.

Courtesy, New Holland Division of Sperry Rand

Fig. 19.9—Silage lends itself to mechanized feeding.

time exceeds 24 hours, top layers of the silage will spoil where exposed to the air.

Length of cut

Short-cut forage (¼ to ½ inch) is preferable to long-cut forage since it packs tighter and more densely, and excludes more air. The drier the forage, the more important the shortness of cut, and in all cases sharp knives are imperative.

Compaction

The forage should be compressed sufficiently in the silo to exclude all air from the silage mass. Since the pressure exerted by silage is great, the walls of upright silos have been known to collapse where excessively wet forage has been stored or where the walls have not been sufficiently reinforced.

"Grass" silage types

Grass—all common perennial or biennial grasses.
Legume—alfalfa, the clovers, soybeans, and lespedeza.
Legume-grass—60 percent or less forage grass.
Sudangrass—whole plants.
Cereal—oats, wheat, rye, etc.
Sorghum—whole plants.
Miscellaneous—soybean-sorghum, pea vine, sunflower, cannery wastes.

CHAPTER 20

Diseases and Insects Take Their Toll

Unfortunately, pasture grasses and legumes, like other plants, are attacked by diseases. So-called "running out" stands can often be traced to attacks by diseases which build up progressively from one year to the next. All parts of the plant, and plants in all stages of growth, are susceptible to attack by one or more of the fungi, bacteria, or viruses.

Certain disease-producing organisms live for long periods in the soil; others live in residues of plants killed by the diseases; still others remain within diseased, living plants and thus become a source of infection for nearby plants.

Activities Which Involve Approved Practices

1. Preventing disease from getting a start.
2. Identifying and controlling diseases common to most grasses.
3. Identifying and controlling diseases of orchardgrass.
4. Identifying and controlling diseases of bromegrass.
5. Controlling eyespot on timothy.
6. Controlling net blotch on fescues.
7. Controlling diseases of bluegrass.
8. Controlling diseases of sudangrass.
9. Controlling seed decay losses.
10. Controlling diseases of legumes.
11. Integrated insect control systems.

No attempt is made in this chapter to discuss all of the diseases affecting pasture plants. Only some of the more common are included.

1. Preventing Disease from Getting a Start

Resistant varieties offer the best means of preventing diseases. Many of the newer varieties possess some resistance to the most serious diseases. There is no substitute for good management practices, including crop rotation, to help reduce and keep diseases under control. Likewise, it is well known that plants growing on fertile soils are more able to withstand attacks by pathogens than slower-growing plants on poor soils.

In the case of foliage diseases, partial control may be obtained by grazing at a stage before the disease normally attacks the plant.

2. Identifying and Controlling Diseases Common to Most Grasses

Ground stripe

This disease is particularly severe on timothy, orchardgrass, tall oatgrass, redtop, and bluegrass. Although present throughout the growing season, it is most evident during mid- and late summer when the leaves and culms are maturing. Severe attacks cause pre-withering and dying of the leaves. The quality of the herbage is lowered, and attacked plants are weakened.

The fungi produce circular to elongated water-soaked spots on the leaves which later turn a deep, dull gray. Still older lesions appear as narrow purple to brown streaks, usually with ashy gray to almost white centers. These lesions may be ¼ inch to several inches long and appear singly or as groups on the leaves and sheaths. Within the lighter center of each lesion appear black dots in parallel rows.

Approved methods of control involve the removal or even burning of the dead grass to reduce overwintering of the fungi. The ultimate solution undoubtedly lies in the development of resistant varieties.

Smooth bromegrass is one of the few pasture grasses resistant to this disease.

Anthracnose

This is one of the most widely distributed diseases on pasture plants. It is particularly noticeable in late summer and fall on many grasses as they mature. When conditions are favorable for development of the disease, the fungus attacks seedlings, causing stunting, wilting, and even death. Although few seedlings may be killed, the damage to the root systems is sufficient to reduce yields. In older plants the fungus may spread to the crowns and roots of perennial grasses and will cause stands to die out in the second or third year.

Approved control measures include:

1. Crop rotation.
2. Avoiding a sequence of closely related crops.
3. Maintenance of high soil fertility.
4. Complete turning under of plant residues.
5. Use of resistant varieties.

Powdery mildew

Although this disease is most destructive on wheat, barley, and other small grains, it also attacks many pasture grasses. The disease is most severe in seed fields. The dusty, mildewy area, first white, and later gray or buff, may completely cover the leaf, causing it to yellow or brown. The fungus overwinters on dead leaves of winter-growing grasses. Disease symptoms are evident in late spring and with cooler weather in the fall may appear again.

While powdery mildew may be checked with fungicides, this is seldom practical under field conditions.

Ergot

This is a worldwide disease. It attacks the seed. Therefore, it is of importance under pasture conditions only when seeds develop. It is of first importance since it affects seed production, which in turn influences the availability of seed for pasture purposes. Approved practices for keeping this disease under control include crop rotation and mowing of roadside grasses before blooming stage to reduce infection.

Rust

There are many rusts that attack pasture grasses. These include stem rust, crown rust, orange leaf rust, and yellow leaf rust. The leaf rusts are the ones that do the most damage, since they attack the pasture plant at the grazing stage. Rusts can be controlled only by selecting resistant varieties.

Smuts

There are at least three smuts that cause considerable damage to pasture grasses. These include stripe smut, which attacks timothy, redtop, Kentucky bluegrass, and orchardgrass; flag smut, which occurs on bluegrass, timothy, and redtop; and black loose smut, which infects tall oatgrass. Seedlings become infected presumably from seed contamination or from spores in the soil. Since the mycelium is systemic, it can overwinter in the crown of perennial grasses.

Treatment of contaminated seed with fungicides reduces the primary infection.

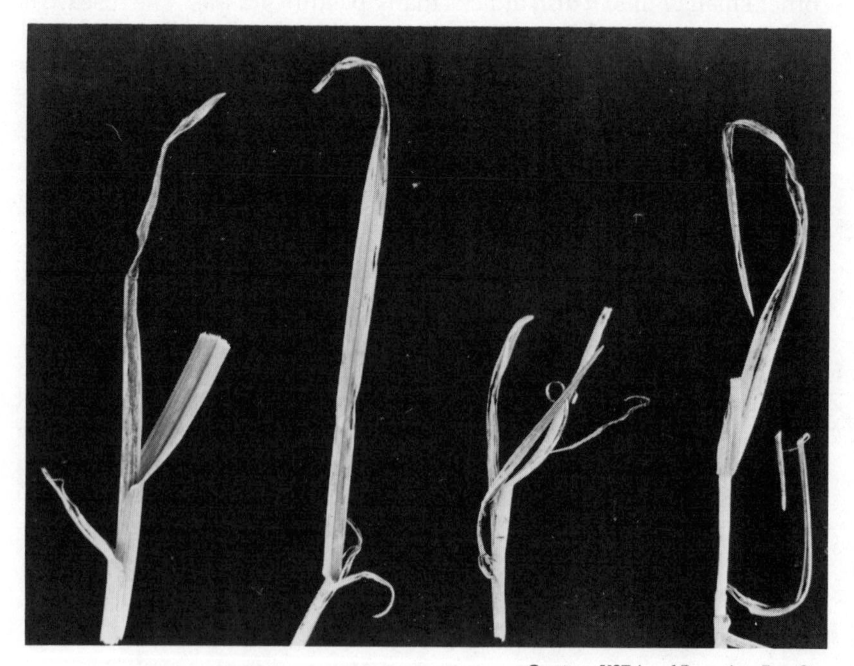

Courtesy, USDA and Penn. Agr. Exp. Sta.

Fig. 20.1—Stripe smut of Kentucky bluegrass.

3. Identifying and Controlling Diseases of Orchardgrass

Purple leaf spot

This disease is of economic importance on orchardgrass being grazed since the infected leaves wilt and die. Infection occurs from spores overwintering in dead stems and leaves.

Breeding for resistance is apparently the only practical means of controlling this disease. Plant breeders have isolated resistant lines.

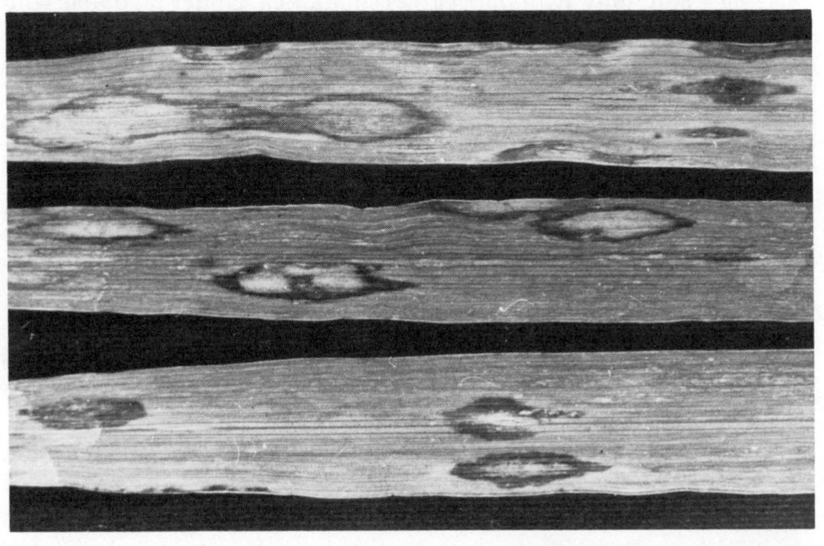

Courtesy, USDA and Penn. Agr. Exp. Sta.

Fig. 20.2—Leaf scald of bromegrass.

Scald

This is another disease that attacks orchardgrass at a time when the plant is growing rapidly. Lesions on the leaves are oval to elongated, light tan to white, and they are surrounded by a tan to reddish-brown border. Weakened leaves may break off at the point of infection. Plant breeders think it may be possible to develop resistant varieties should this disease become a serious menace.

4. Identifying and Controlling Diseases of Bromegrass

The three diseases most prevalent on bromegrass are leaf scald *(Rhynchosporium secalis)*, brown spot *(Helminthosporium bromi)*, and bacterial blight *(Speudomomas coronafaciens var. atropurpurea)*.

Leaf scald

This appears as a dark, bluish-gray, water-soaked blotching that becomes light gray with dark brown markings. Under favorable conditions the lesions enlarge and elongate until large areas of the leaf are affected and killed. Leaf scald results in premature death of leaves and reduces the quality of the herbage.

The fungus overwinters on dead leaves and old crop residues. Under mild conditions, new lesions may be produced throughout the winter. Spores are produced during cool, moist weather and are carried to healthy plants by wind and rain. The spores germinate at 32° to 82° F. but the optimum temperatures are 65° to 75°. Fortunately, resistant strains of smooth bromegrass are available.

Brown spot

Almost all fields of smooth bromegrass are damaged by brown spot. Infection, as first noticed, consists of small, brown, oblong spots on the leaves which develop first in the spring. This disease develops best during cool, wet weather and then spreads very little during the hot, dry weather of mid-summer. It shows up again in the fall.

Varieties resistant to brown spot are the best means of controlling the disease. Resistant lines are being investigated at many of the experiment stations and may ultimately compose varieties that will reduce loses from this destructive foliar disease.

Bacterial blight

Bacterial blight is most noticeable during periods of warm, humid weather. Earliest symptoms on the leaves consist of small watery spots with brown centers which later enlarge, become somewhat linear, and appear chocolate-colored or almost black. Although this disease is not so serious as some on smooth bromegrass, it is capable of causing severe damage under certain conditions.

5. Controlling Eyespot on Timothy

The most common foliar disease on timothy is eyespot, the scientific name being *Heterosporium phlei*. This disease occurs primarily in the East and Midwest and to a limited extent in the Far West. Lesions may be so numerous that the leaves are killed prematurely. Although eyespot is most prevalent during summer months, it has been found on green leaves in almost any season.

Courtesy, USDA and Penn. Agr. Exp. Sta.

Fig. 20.3—Eyespot of timothy.

Eyespot occurs on the leaves as small light-colored lesions with a narrow violet border which fades to brown. When the lesions are abundant the intervening tissue frequently becomes yellow.

Spores are capable of germination at temperatures from 37 to 91 degrees F. Unfortunately, no concentrated effort has been put toward the control of this disease. Some of the newer varieties are somewhat resistant to eyespot.

6. Controlling Net Blotch on Fescues

Helminthosporium dictyoibes, or net blotch, is probably the most common foliar disease of tall fescue and meadow fescue. Leaf lesions which are found throughout the growing season range from small, oval, brown spots to extensive brown, net-like streaks which may cause complete withering of the leaves and sometimes death of the plant. The fungus is seed-borne. Research workers have found differences among several strains of fescue, which suggests the possibility of isolating strains resistant to net blotch.

7. Controlling Diseases of Bluegrass

Leaf spot *(Helminthosporium hagans)* and leaf blotch *(Septoria spp.)* are two diseases of economic importance on bluegrass. The fungus disease leaf spot occurs in pastures in the cooler humid parts of North America and is most prevalent in the eastern and central parts of the United States. Thick stands of bluegrass are more seriously affected than mixed stands of grasses. Damage from leaf spot is accelerated by close grazing since the succulent growth is the most susceptible. Therefore, applications of nitrogen fertilizer during summer months favor the development of the disease because of nitrogen stimulating new growth.

The leaf spot appears in the purple-black to reddish-brown lesions that vary up to ½ inch in length. Older spots usually have a light center surrounded by a red or purple margin. Some lesions run the entire width of the leaf, causing it to break over or wither from the tip to the point of attack. Infections occur throughout the growing season from spores produced from older lesions. The spores are spread to healthy leaves by wind and rain. The fungus overwinters in lesions in living leaves or as spores and mycelium on dead leaves.

Further work is necessary to determine the extent of seed infection and whether seed treatment with fungicides is effective.

Moderate grazing helps keep leaf spot under control. Strains of Kentucky bluegrass differ in susceptibility, with Merion bluegrass having resistance.

Although many species of *Septoria* attack grass, two species, *S. macropoda var. septulata* and *S. oudemansii*, are widespread. Both diseases occur during periods of cool, wet weather and are therefore most common in spring and fall. Infection comes from spores that overwinter on dead leaves and old plant parts. The spores germinate over a wide range of temperatures. They are spread to healthy leaves by splashing rain and wind-carried diseased plant parts.

Some reduction in disease can be achieved by removing or plowing under plant residues. Strains of bluegrass differ in susceptibility. Resistant varieties could undoubtedly be developed.

8. Controlling Diseases of Sudangrass

The leaf spot disease *Helminthosporium turcicum* is prevalent on common sudangrass from mid-summer until the plant matures. It is most noticeable during warm, moist weather. It shows as elongated spots, tan to reddish-purple in color, often with a dark margin. Later the lesions increase in size and fade to a gray or straw color, causing the leaves to appear blighted.

Varieties of sudangrass, such as Piper and Tift, have resistance to leaf blight. Newer varieties are more resistant.

9. Controlling Seed Decay Losses

Seed decay diseases are active during wet periods. Among the prevalent soil-borne fungi that attack both grasses and legumes are the species of *Pythium, Fusarium,* and *Rhizoctonia solani.* These diseases cause seed decay in wet weather. Some of the seedings actually die before they emerge. Others wilt rapidly after emerging, a condition often referred to as "damping off." The affected plants that continue to live are pale yellowish-green and grow slowly.

Reducing "damping off" losses

There are several approved practices which help reduce losses from "damping off," "seedling blight," or "root rot." These are:

1. Adequate use of lime and fertilizer.
2. The preparation of a good firm seedbed which enables the seedling to emerge rapidly and vigorously.
3. Treatment of the seed with a fungicide.

10. Controlling Diseases of Legumes

Bacterial wilt in alfalfa

This common disease is favored by an abundance of moisture and is usually most severe in low, poorly drained soil. Plants infected are stunted and yellow. The shortened stems result in a bunchy growth. The leaves are short, chlorotic (non-green), and often cupped. During warm, dry days, plants infected by wilt die rapidly.

The disease can be identified by examining the taproot. When the bark below the ground is peeled, a yellowish or pale brown discoloration of the outer wood, in contrast to the creamy white appearance of healthy wood, characterizes the infection. The causal bacteria enter the plants through wounds produced by winter or mechanical injury.

The approved practices for the control of bacterial wilt include:

1. The planting of resistant varieties. Most new varieties adapted to the humid region are resistant.

Fig. 20.4—Bacterial wilt of alfalfa causes stunting of plant growth. The most severely infected plants are on the left and the least infected plants are on the right.

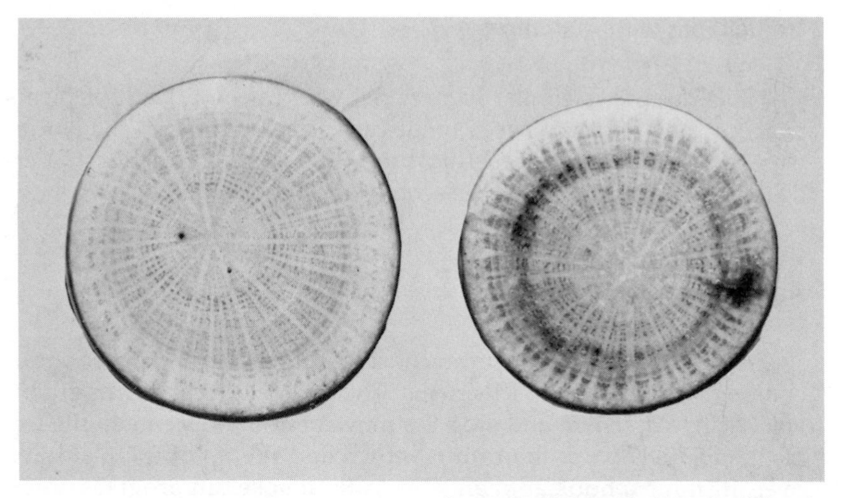

Courtesy, Fred R. Jones, The University of Wisconsin

Fig. 20.5—Cross-section of healthy and bacterial-wilt infected alfalfa roots.

2. Maintenance of high soil fertility, particularly by applications of phosphate and potash.
3. Prevention of mechanical injury to crowns and roots.
4. Not reseeding old alfalfa fields until all roots have completely rotted.
5. Not seeding fields that receive surface drainage from infected areas.

Phytophthora root rot of alfalfa

This disease is prevalent in the humid region of the U.S. and is associated with wet soils. The fungus spores are swimming spores, thus high soil moisture is necessary for phythophthora root rot to spread. Phytophthora root rot attacks all ages of plant life, from seedlings to older, mature plants. Selecting well drained soils for alfalfa production reduces phytophthora root rot infections. Many high-yielding varieties are resistant to phytophthora root rot.

The taproot of an infected plant develops a darkened lesion. The lesion may rot the lower portion of root off, then progress upward on the root.

Verticillium wilt of alfalfa

This disease is similar to bacterial wilt, causing death of plants at about three years of age. The disease appeared first in the Northwest U.S. and has moved into the Midwest and Northeast. Verticillium wilt causes leaves to become blanched and curled.

Anthracnose of alfalfa

Alfalfa is attacked by anthracnose during moist, warm periods of summer. The disease attacks stems. The dark, elliptical lesions gradually encircle the stem and stop the movement of nutrients to the top of the plant. The top then turns brown and the tip of the plant tips over, forming a hook appearance. Anthracnose can progress down into the crown and kill the plant. Many varieties have moderate to high levels of resistance.

Alfalfa leaf spot

Common leaf spot is perhaps the most destructive foliage disease of alfalfa. It's not a killer but the plants lose foliage, which reduces the quality of the forage and lowers the vigor of the stand. The disease is characterized by small, circular brown spots approximately 1/16 inch in diameter which develop on the leaflet. When the spots are fully developed the centers become thickened and a tiny, light-brown, cup-shaped lesion forms.

The fungus overwinters on dead, undecayed leaves. It is believed that the disease is not seed-borne. Most varieties have moderate to good resistance to the disease. Plant breeders have discovered highly resistant lines and are incorporating resistance into new varieties.

Downy mildew

Downy mildew attacks alfalfa during wet or humid weather, particularly in the spring. It occurs commonly on young plants, weakening them and causing defoliation. The disease occurs first on the upper leaves. These infected leaves become light green in color, and later yellow, twisted, and rolled. A fine, gray, mold-like growth appears on the underside of the leaves. Numerous spores produced on diseased leaves are spread by the wind.

Anthracnose of red clover

The disease most commonly associated with red clover is anthracnose. There are two strains of this destructive disease—northern anthracnose, which occurs in the cooler parts of the country, and southern anthracnose, which occurs primarily in the southern states.

Losses of 50 percent or more of the crop have been reported in some instances. Forage production and quality as well as seed yields are seriously affected by heavy attacks. Symptoms are confined chiefly to petioles, leaves, and stems. Early symptoms on petioles consist of dark brown to black spots which become light-colored with dark margins. Plants in fields heavily attacked look as though they have been scorched by fire. Presumably the fungus overwinters in dormant tissue of the crown. Wind and splashy rains carry spores from lesions to healthy plants.

Courtesy, USDA and Penn. Agr. Exp. Sta.

Fig. 20.6—Target spots of red clover.

Courtesy, USDA and Penn. Agr. Exp. Sta.

Fig. 20.7—Powdery mildew of red clover.

Courtesy, USDA and Penn. Agr. Exp. Sta.

Fig. 20.8—Pepper spot on Ladino clover. No practical control measures are known.

While it is impossible to distinguish between foliar symptoms of northern and southern anthracnose, the latter disease frequently attacks the upper part of the taproot and crown. Diseased crowns infected with southern anthracnose often tend to be brittle, and the stems break off readily at the soil level.

Resistant varieties offer the best means of control.

Sclerotinia attacks many clovers

Perhaps one of the most damaging diseases on clovers is sclerotinia crown and stem rot. This disease causes the crown and stem to rot in early winter and early spring. It also attacks alfalfa and birdsfoot trefoil.

Greatest damage occurs in mild winters or under snow cover. Plants of all ages are susceptible. As a plant part dies, a white, fluffy mass grows over the diseased area forming hard, black sclerotia which may adhere to the surface of or be imbedded in the stem, crown, or root. They look very much like mouse feces.

The fungus can remain in the soil for years. The longevity of the sclerotia makes control difficult. The following approved practices, however, are recommended:

1. Planting seed free of sclerotia.
2. Deep-plowing to bury sclerotia beyond their capacity to develop fruiting bodies.
3. Following long rotations.

11. Integrated Insect Control Systems

Many insects attack pastures and forage crops. Among the common insects are grasshoppers, army worms, alfalfa weevils, Egyptian alfalfa weevils, aphids, Lygus bugs, caterpillars, webworms, mites, leaf miners, leafhoppers, thrips, sowbugs, and slugs.

Most, if not all, of the common insect pests can be controlled by chemicals. However, chemical pesticides are poisonous compounds and must be used with caution. Many chemical pesiticides are not cleared by the federal government for use on pasture or hay crops for one reason or another. Others are cleared with restrictions as to rate and time of usage.

As effective as insecticides can be, they should be used as only one practice in an integrated system which includes destruction of host plants, timely clipping, rotation of crops, and encouragement of predators. Officially approved insecticides should be applied strictly in line with directions on the labels.

Since new chemicals are constantly appearing on the market or have recently been cleared for use on forages, and since others may be removed from the recommended list, the reader is referred to the local vocational agricultural teacher, county farm advisor or agent, or other authorities for recommendations as to which insecticides to use for his or her particular difficulties.

Regardless of which chemical is used, the insecticide should be handled with caution. Recommendations on the use and storage as well as handling of insecticides are published in such bulletins as Circular 554 of the California Agricultural Extension Service. The circular makes these recommendations:

> Pesticides are poisonous and always should be used with caution. The following suggestions for the use and handling of pesticides will help minimize the likelihood of injury, from exposure to such chemicals, to man, animals and crops, other than the pest species to be destroyed.
>
> **Always** read all precautionary labeling directions before using sprays or dusts and follow them exactly. Notice warnings and cautions before opening the container. Repeat the process every time, no matter how often you use a pesticide, or how familiar you think you are with the directions. Apply material only in amounts and at times specified.
>
> **Personal safety**—Avoid inhaling sprays or dusts. When directed on the label, wear protective clothing and a proper mask.
>
> **Never** smoke, eat, or chew while spraying or dusting.
>
> **Application**—Do not use the mouth to siphon liquids from containers or to blow out clogged lines, nozzles, etc.
>
> Do not spray with leaking hoses or connections.
>
> Do not work in the drift of a spray or dust.
>
> Confine chemicals to the property being treated and avoid drift by stopping treatment if the weather conditions are not favorable.
>
> Do not apply pesticides over fish ponds, canals, streams, or lakes, and do not apply them to fields being irrigated if drain water runs off field.
>
> Observe label directions and follow recommendations in order to keep the residue on edible portions of plants within the limits permitted by law.
>
> **Low Volume Sprays**—Although low volume sprays have

proved effective in certain regions with large acreages of uniform crops (cotton in South, forest, cereal and grasslands), their application to the diversified agriculture in California is not appropriate. Research has indicated that LV sprays and in particular LV technical material applications cause increased drift residues and present unacceptable hazards to honey bees and other beneficial insects. For these reasons, application rates of less than 5 gallons per acre are not recommended.

Decontamination—Should pesticides be accidentally spilled on the skin or clothing remove contaminated clothing *immediately* and wash the contaminated skin thoroughly.

Bathe and change to clean clothing after spraying or dusting. If it is not possible to bathe, wash hands and face thoroughly and change clothes. Also wash clothing each day before re-use.

In case of illness—If symptoms of illness occur during or shortly after dusting or spraying, call a physician or get the patient to a hospital *immediately*.

Disposal of empty containers—Always dispose of empty containers so that they pose no hazard to humans, animals, valuable plants or wildlife. See your County Agricultural Commissioner's Office.

Storage—Keep pesticides out of reach of children, pets, irresponsible persons, and livestock. They should be stored outside the house, away from food and feed, and under lock and key.

Always store sprays and dusts in their original containers and keep them tightly closed. *Never* keep them in anything but the original container.

Working in treated areas—If pickers or others are working in trees or crops with heavy foliage, such as peaches, olives, citrus, etc., or in grapes that have been treated with highly toxic compounds, be sure the recommended interval between the treatment and entrance into the treated area is observed. These workers should follow the same precautions as given for the applicators in regard to changing clothing, wearing protective clothing, eating or smoking, and bathing. If a worker becomes ill while working under these conditions, call a physician immediately.

CHAPTER 21

Controlling Weeds in Pastures

Weeds are objectionable in pastures for the following reasons:

1. They lower the feeding value of the forage.
2. Weeds rob the desirable species of plant food and moisture.
3. Weeds shade and crowd out the more nutritious legumes and grasses.
4. Weeds can and do lower the value of livestock products produced as a result of pastures. Examples are wild onion or garlic flavors in milk and cockleburs in wool.
5. Certain weeds are poisonous and, when eaten, can cause death to livestock.
6. Weed growth gives pastures and farmsteads an unsightly appearance.

Activities Which Involve Approved Practices

1. Eliminating weeds prior to seeding.
2. Mowing to control weeds.
3. Controlling weeds chemically.
4. Reseeding.
5. Reseeding no-till.

1. Eliminating Weeds Prior to Seeding

Prepare a good seedbed

The efficient farmer always prepares a good seedbed. Its prepara-

Courtesy, Soil Conservation Service

Fig. 21.1—A weed-free pasture supplies ample nutrients for livestock.

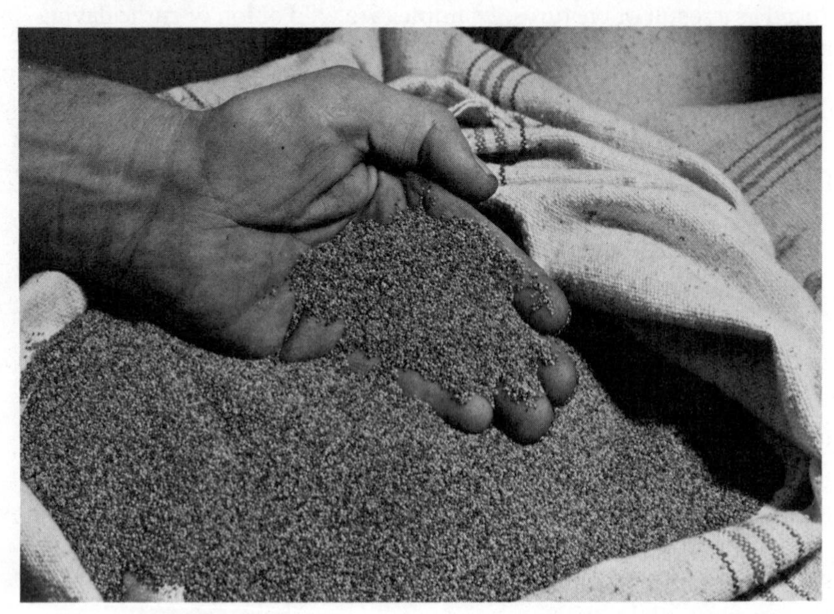

Courtesy, U.S. Bureau of Reclamation

Fig. 21.2—Weed-free seed is the most economical in the long run.

tion includes the killing of all old vegetation if the pasture is being completely reseeded. However, a completely weed-free seedbed is also necessary if the pasture is being established on new land. If weeds are present, tillage operations will destroy large numbers of them. Also, if the land is prepared by first plowing, this operation will bury many of the seeds so deeply that they cannot germinate.

In addition to taking every precaution to eliminate volunteer weeds, the good farmer uses a high-quality crop seed—seed not only of adapted legumes and grasses but also free of noxious and undesirable weeds. Thus, the first step in controlling weeds starts at the time the pasture is established. From then on, other approved practices help keep them under control.

2. Mowing to Control Weeds

Once a pasture is established, systematic mowing helps keep down weeds for the following reasons:

1. Mowing kills annual weeds if the pasture is mowed before these weeds go to seed.
2. Mowing prevents perennial weeds from going to seed, thus restricting their spread.
3. Some weeds, mowed when they are young, are eaten and enjoyed by livestock.
4. Regular mowing prevents weeds from shading desirable grasses and legumes.
5. Mowing, in addition to keeping weeds under control, maintains desirable plants at a uniform height and maturity so that the entire field is palatable to the grazing livestock. This eliminates spot grazing in the pasture.

Mowing has many advantages

Mowing pastures to control weeds is simple and effective, and, you might say, one of the more pleasant chores on the farm from the farmer's standpoint. Here are the advantages of mowing as a means of weed control:

1. A mower is standard equipment on virtually all farms today.
2. Danger of poisoning animals is eliminated.

3. Mowing leaves traces of dry feed in the pasture. The animals eat these clippings. They help prevent bloat.

4. Mowing does not harm the desirable grasses and legumes. In fact, mowing often stimulates their growth. Occasionally in chemical treatment some of the good species are hurt as well as the bad ones.

5. There are no chemicals or spray equipment to buy.

Mowing has disadvantages too

Mowing, however, when compared with weed control by chemical treatment, does have disadvantages. They are:

1. Mowing does not kill perennials, but merely keeps them from producing seed.

2. It normally takes longer to mow a field than to cover it with a modern, long-boom spray rig.

3. Sprayers cover fence rows and weeds next to stones, areas where a mower cannot reach conveniently.

3. Contolling Weeds Chemically

We are living in a chemical age. Each year there are more new improved herbicides which are selective in their killing action. The use of weed killers (which are herbicides) provides the farmer with an efficient and economical way of destroying unwanted plants.

Follow manufacturer's directions

It is essential, however, that the directions of the manufacturer be followed in using any weed killer.

The reader must realize that herbicides are classified as *selective* and *general contact* chemical weed killers. The former kill some plants without harming others, while the latter destroy all types of vegetation, desirable or undesirable.

You can kill most annual weeds with one timely application of chemical at the recommended rate, although a second spraying is sometimes necessary to control late-emerging plants. Early fall, fall, or early spring treatments are equally effective in the control of many annuals and biennials. Weeds are most easily controlled by herbicides when they are growing actively under conditions of good soil moisture, high fertility, and warm temperatures. Effective control of many troublesome broadleaf weeds may require early summer applications of approved herbicides for three or four successive years or longer if persistent weeds like vervain or ironweed are present. Thistles, annual ragweeds, asters, gumweeds, and the like generally succumb after two consecutive years of application.

Herbicides, like insecticides, are being critically scrutinized by governmental agencies. It is recommended that the reader check with qualified authorities before using a herbicide about which he or she may have questions.

Advantages of chemical control

1. Chemical costs are usually relatively low, often less than the cost of mowing.
2. Herbicides can be used to destroy weeds in fence rows and other areas inaccessible to a mower.
3. On land too steep or rough for mowing or other mechanical treatment, sprays can be applied by airplane.
4. Increases in forage production can be expected after the first year of treatment.
5. Commercial application of herbicides is available in most areas, making it unnecessary to invest in special equipment and eliminating interference with other farm work.
6. Perennials do not escape the action of herbicides.

Disadvantages of using chemicals

1. Spraying must be done when temperature and growth conditions are within certain limits.
2. Certain herbicides are poisonous to livestock, which must be removed from the field when treatment is made.
3. Special equipment may be needed, and critical periods of application may come when one is busy with other farm tasks.
4. Weedy grasses are not affected by commonly used weed sprays.

Courtesy, Dow Chemical Co.

Fig. 21.3—A host of new weed killers to apply as sprays are available for controlling weeds.

5. Legumes, classed as broad-leaved, will be injured, at least temporarily, by most herbicides used to control broadleaf weeds.

4. Reseeding

Plowing, followed by reseeding with desirable species, is a good way to handle a weedy pasture if it can be farmed for at least one year to small grain, sorghum, sudangrass, or similar crops on the plowed-up pasture. It may be desirable to use some form of cultivation prior to seeding in the stubble to kill weeds and volunteer growth. In any event, the seedbed must be made firm by rolling following any tillage operation for seedbed preparations.

Advantages of reseeding

1. A thorough and effective job of weed control is accomplished.
2. Desirable grass and legume species replace undesirable plants.
3. A greater production results from the renovated pasture.

5. Reseeding No-Till

Highly erosive soils are especially well suited to no-till reseeding of pastures. No-till reseeding objectives include weed control, incorporation of higher-yielding and higher-nutritive-value species, and increased fertility.

There are various techniques for successful reseedings by no-till methods. Some methods employ techniques to selectively eliminate some species, such as broadleaf weeds, and leave a preferred species in place. Most usually, broadleaf weeds are killed, the grasses are subdued by overgrazing and herbicide, and a legume or mixture of legumes is interseeded into the grass.

In the arid regions of the Western Plains, competition from the native grass is eliminated by a strip tillage tool and seeder combination. A competitive legume, usually alfalfa, is seeded into the 4-6 inch wide tilled strip (shallow furrow). Under the dryland conditions, the grasses do not re-enter the tilled strip before the legumes become well established and can compete with the native grasses.

Humid-region reseeding of pastures no-till usually requires a her-

Fig. 21.4—No-till pasture renovation seeders place seeds of improved species in contact with the soil.

bicide to set back or kill the grass vegetation. The legume or legume mixture (which may also include an improved grass species) can be seeded with a no-till seeder. No-till seeders may do a small amount of tillage (1-2 inch wide strips) or simply prepare a slot cut by a rolling coulter. The seed is placed in these minutely prepared seedbeds.

Manage properly

The effectiveness of any weed control program depends upon management practices of stocking rates, proper seasonal use for different grasses and legumes, moisture management, and lime and fertilizer use. Good pasture management should mean pastures free of weeds.

Grasses are favored and weeds discouraged by:

1. Avoiding overstocking of the pasture.
2. Occasionally giving the pasture a rest period.
3. Reducing run-off and evaporation of moisture by leaving some vegetative growth uneaten (3 inches for short grass and up to 7 inches on taller species).
4. Applying lime and fertilizer to the pastures as needed (shown by soil tests).
5. Grazing the grasses and legumes during the times when they are making their best growths (set up a pasture calendar).

CHAPTER 22

Productive Pastures and Our Environment

In recent years, there has been growing concern as to what people are doing to their environment. Ecologists and environmentalists are vocal, and many of them point the finger at agriculture as a major contributor to the deterioration of the environment. They reason that soil erosion and leaching of nitrates and other nutrient materials from organic matter, manure, and fertilizers pollute our streams and underground waters.

While there is certainly a need for careful consideration and the introduction of new methods and materials under some conditions, many of those most vocal on the subject are rightly concerned but lack a solid technical knowledge in any area of ecology. The protection of the environment has in many instances become an emotional crusade.

Ways in which productive pastures contribute to an improved ecology include:

1. Recycling carbon dioxide to generate oxygen.
2. Removing certain pollutants.
3. Lessening soil and wind erosion.
4. Filtering out impurities before they reach the ground water.
5. Beautifying and utilizing land efficiently in open-space developments.
6. Providing space and cover for wildlife.

1. Recycling Carbon Dioxide to Generate Oxygen

The air over an acre of farm land contains about 37,000 tons of nitrogen, 9,500 tons of oxygen, and 21 tons of carbon dioxide. When a green plant grows in sunlight, with adequate nutrients and water, it converts about 260 units of carbon dioxide to 180 units of dry matter and releases about 190 units of oxygen to the atmosphere. The more actively growing plants such as those found in productive pastures generate the most oxygen and, of course, use up the most carbon dioxide. Table 22.1 shows the relative quantities of carbon dioxide used and oxygen generated by different forage plants.

Table 22.1

The Utilization of Carbon Dioxide and Generation of Oxygen by Select Forage Plants

Forage	Yield (Air-dried)	Carbon Dioxide Used	Oxygen Generated
	(tons/acre)		
Alfalfa	8.0	10.40	6.80
Grass-legume hay	3.5	4.55	2.88
Bluegrass	2.5	3.25	2.13
Bromegrass	7.0	5.20	3.40
Bermudagrass	7.0	4.10	5.95
Orchardgrass	5.0	6.50	4.25

Plants utilize carbon dioxide and give off oxygen through the process called "photosynthesis." Plants on the earth and in the sea recycle the carbon dioxide generated by humans, animals, and even industries by manufacturing dry matter and releasing oxygen through the photosynthetic process. The photosynthesis equation is:

$$6\,CO_2 + 12\,H_2O \rightarrow C_6H_{12}O_6 + 6\,O_2 + 6\,H_2O$$

(6 carbon dioxide + 12 water) → (1 carbohydrate + 6 oxygen + 6 water)

An average adult requires about 2 pounds of oxygen per day. A person at hard physical labor utilizes much more, and an athlete uses up tremendous quantities of oxygen. All animal life, of course, demands oxygen, and our industries generating energy utilize high amounts of oxygen.

Water is released during photosynthesis. This release of moisture

Courtesy, USDA

Fig. 22.1—Productive pastures recycle large quantities of carbon dioxide.

has a cooling effect on the atmosphere. Dr. John Carew of Michigan State University says, "On a single day, an acre of grass will lose about 2,400 gallons of water through transpiration and evaporation."

In the process of transpiration, grass, in contrast to hard-surfaced roads and barren land, tempers the heat from the sun and provides natural cooling. Grass is the most important air conditioner because of its extent and continued growth even when it is mowed and grazed every day.

The point to be remembered is this: The more productive the pasture—the faster the growth and the more the growth—the greater the contribution to the improved environment through greater utilization of carbon dioxide and the greater generation of oxygen.

2. Removing Certain Pollutants

L. E. Craker, a researcher at the University of Massachusetts, demonstrated that soil has the power to remove certain pollutants other than carbon dioxide from the air. Scientist Craker did his research at the University of Massachusetts Suburban Experiment Station in Waltham in cooperation with scientists at the USDA's Plant Air Pollution Laboratory, Beltsville, Maryland, and the Army's Plant Science Laboratory at Fort Detrick, Maryland.

The work began at Beltsville with the air pollutant ethylene and continued at Waltham with other air pollutants, namely sulfur dioxide and nitrogen dioxide—major pollutants caused by combustion. Dr. Craker says:

> "We found that when we placed sulfur dioxide and nitrogen dioxide in containers with small amounts of soil the pollutant disappeared at a very rapid rate. For example, within 15 minutes after we added approximately 100 parts per million of sulfur dioxide to a container with 250 grams of soil only 8% of the pollutant remained." When the same tests were performed with soil sterilized by heat, the pollution removal was slowed down somewht but not halted. The results suggest that while microbial action may play some role in removal, a major portion of the pollutants are removed by the soil itself in some chemical fashion.

Productive pastures build productive soils by increasing the soil's tilth and natural fertility, which adds up to a greater contribution to clean air.

3. Lessening Soil and Wind Erosion

The major contributor to pollution of surface waters is sediment. In the United States, about half of all sediment comes from agricultural lands. It is estimated that 4 billion tons of sediment are carried into waterways and lakes in the United States annually from surface erosion. For example, the Mississippi River each year carries about 500 million tons of sediment to the Gulf of Mexico.

Permanent vegetative covers, such as productive pastures, provide the most effective means of reducing soil erosion. The foliage intercepts the "pounding" action of raindrops, reducing the force with which they hit the soil. The energy of the rain is dissipated, and the water trickles down the foilage to the surface of the soil. Much more of the water soaks into the soil in this way, and run-off is materially reduced.

In addition, the velocity of the water that does run off is slowed down by the mechanical obstruction offered by the forage and sur-

Courtesy, USDA

Fig. 22.2—Productive pastures are excellent "protectors" against water erosion.

face residues. This reduces the erosive power of the water, and surface soil erosion is cut substantially.

Productive pastures are associated with massive root systems through the soil profile. These roots hold the soil particles in place and thus contribute to reduced loss of soil. The roots also improve soil tilth and increased water infiltration at the surface layer.

Data showing the degree to which good sods control soil erosion are presented in Table 22.2.

4. Filtering Out Impurities Before They Reach the Ground Water

The deep, massive root system associated with productive pastures not only hold the soil in place and slow down run-off but also act as a "filter wick" for soluble impurities that might otherwise leach down to the ground water.

Table 22.2

Soil Losses from Surface Run-off As Influenced by Types of Cover[1]

Location	Soil Type	Slope	Cover	Soil Loss (tons/acre)
Bethany, Missouri	Shelby silt loam	8.0	Bare, uncultivated soil	112.48
Bethany, Missouri	Shelby silt loam	8.0	Continuous corn	61.16
Bethany, Missouri	Shelby silt loam	8.0	Continuous bluegrass-timothy	0.36
Bethany, Missouri	Shelby silt loam	8.0	Continuous alfalfa	0.22
Tyler, Texas	Keiven fine sandy loam	8.75	Bare, uncultivated soil	12.20
Tyler, Texas	Keiven fine sandy loam	8.75	Continuous cotton	19.06
Tyler, Texas	Keiven fine sandy loam	8.75	Bermudagrass	0.20

[1]Soil with an active growing cover is well-protected against action of winds. Wind erosion is effectively controlled.

Actively growing plants take up soil moisture in very large amounts. It has already been shown in this chapter that an acre of grass will lose about 2,400 gallons of water through transpiration and evaporation. This water enters the plant through the roots. As water moves to the roots it carries with it water-soluble materials such as nitrates. Most of these soluble materials also are absorbed by the growing roots as they come in contact with the root system.

During periods of excess moisture, the water and soluble materials move downward. Rapidly growing and massive roots—the kind of root systems found in productive pastures—intercept and filter out the soluble materials as the roots intercept the downward flow of moisture carrying the soluble materials. As a result, most of the soluble materials the water carries never reach the ground water table.

In periods of little or no rainfall or added water through irriga-

Fig. 22.3—Productive pastures have extensive roots that aid in purifying ground water.

tion, soil moisture moves upward toward the soil's surface. Water that may have percolated beyond actively growing roots "wicks" upward again, carrying dissolved materials. This upward movement is intercepted by the roots. The water is utilized, and soluble nutrients such as nitrates are again removed from the soil solution.

The soil has been referred to as a "live filter." Most of the soil microorganisms also thrive in the active root zone and consume soluble nutrients as they multiply.

The root system associated with plants of productive pastures—grasses and legumes—is extensive, and therefore productive pastures make a material contribution in keeping the ground water free of soluble contaminants.

5. Beautifying and Utilizing Land Efficiently in Open-Space Developments

Cluster developments with accompanying open space are receiving considerable attention. Many people like to be as close to nature as possible while they are also enjoying the everyday con-

Courtesy, USDA

Fig. 22.4—Pastures and housing can complement each other.

veniences associated with urban living. The home-cluster open-space concept provides both of these elements.

If considerable open space can be kept in productive agriculture without spoiling the beauty of the area, then unit housing costs should be considerably less. Productive pastures are ideal for such use. The green fields are attractive; grazing cattle also add to the scenic beauty. Other contributions to our environment already mentioned in this chapter are extra dividends.

6. Providing Space and Cover for Wildlife

Productive pastures attract birds and small animals. It is also not uncommon to see deer—especially near dusk—grazing on tender, succulent pastures.

Migrating waterfowl frequently interrupt their flights for grazing flood-irrigated pastures. Other animals frequent pastures at all times. Small game seek out pastures for nesting spots away from the dangers of heavily traversed areas.

The combination of lush growing grasses and plants interpopulated with domestic and wild animal life is indeed a step toward perfect ecology.

CHAPTER 23

Pasture Demonstrations and Contests

The adage "Seeing is believing" well applies to pasture improvement. If the farmer sees the many advantages improved pastures have over unimproved ones, and is convinced, he or she is more likely to put into immediate effect approved practices on part or all of the pasture acreage.

Activities Which Involve Approved Practices

1. Planning pasture demonstrations.
2. Creating community interest.
3. Juding hay quality.
4. Judging silage quality.

1. Planning Pasture Demonstrations

The simplest pasture demonstration is merely a narrow strip selected out of a large pasture. This strip is the site of the approved practice, seeding of adapted legumes and grasses, fertilization, clipping, weed control, or a combination of many of the approved practices. Thus, visiting farmers can compare at a glance the improvements as demonstrated on the improved strip and the untouched pasture. The county agent can supply necessary information and tell where to obtain seed.

Long, narrow strips are preferable to wide, square blocks in setting up demonstrations of this nature because the strip may cut across many soil variations in the length of the field and give more representative conditions.

Prevent grazing

One of the difficulties of conducting pasture demonstrations is the fact that the evidence often is "eaten up." The improved strip is usually more heavily grazed by the animals than the area not treated and unless the farmer is careful the treated area may give the appearance of providing actually less grazing than the non-treated area.

Cage small areas

With small animals, it is often possible to use small cages which protect small areas from being grazed by the animals. One or more of these cages are placed on the area given the approved practice treatment, and the growth is visually compared with that under similar cages on the non-treated area.

When pasturing larger animals, it is usually better to fence in a small area running across both the treated and non-treated section. This system affords the farmer an opportunity to observe the difference in growth associated with the approved practices.

Of course, the best way of seeing what pasture improvement can mean is to treat an entire field, or half a field, fencing it off and then comparing it with an adjoining area. Care needs to be taken in such demonstrations that the soils, drainage, fertility, and position to the sun are as near identical as possible. Without this care, the observed superiority of one treatment may really be a response to an unrelated factor such as drainage mentioned above. Often growth alone is not indicative of the total benefits to be drived from pasture improvement, since the treated grass is frequently more nutritious than the herbage from non-improved areas.

2. Creating Community Interest

Use contests

Pasture contests have been used successfully in every section of the

United States. Their use is to be encouraged. These contests, usually held among farmers and based on all-around, approved pasture management practices, have done an excellent job of acquainting the public with the merits of good pasture management, and what such practices can mean to an agricultural community.

Publicize results

Usually the contests are a form of community activity. Farmers compete among themselves, sometimes for prizes, sometimes for high score and public recognition. The farm radio, television, and press have done a splendid job publicizing benefits to farmers of using approved pasture management practices, and how resulting profits can increase the purchasing power and overall prosperity of a particular community.

The "New England Green Pasture Forage Program"

All New England states competed in this well known pasture contest. Educational agencies pooled their efforts in directing the program, and more than 200 commercial organizations contributed to its support. The county agents functioned as "spark plugs" to make the program work.

The contest started in 1948, when Charles M. Dale, then governor of New Hampshire, challenged the other governors of New England states to produce a pasture better than New Hampshire's finest.

The score card can serve as a guide

The score card used in the New England pasture contest is reproduced in Table 23.1. This score card has been modified from time to time and has been used as a basis for developing score cards for other areas.

3. Judging Hay Quality

Putting up high-quality hay is a complementary practice in the production of good-quality pastures.

Table 23.1

Check Sheet and Score Card for New England's Green Pasture-Forage Program

Objective—To produce milk at a lower cost by growing and feeding better quality hay, pasture, and silage.

Name____________________Address____________________County____________

	Points	Score
A. PRODUCTION OF HAY, PASTURE, CORN, AND/OR GRASS SILAGE (35)		
1. Early pasture	5	_____
2. July pasture—hay, grass silage, corn silage	8	_____
3. August pasture—hay, grass silage, corn silage, millet, sudan	14	_____
4. Fall pasture—hay, grass silage, corn silage, aftermath	8	_____
B. QUALITY OF ROUGHAGE (20)		
1. Pasture—hay, grass silage, corn silage	10	_____
2. Early harvesting—proper curing and storage methods	10	_____
C. MANAGEMENT PRACTICES (45)		
1. Wise use of lime, manure, fertilizer	10	_____
2. Clipping, rotation, access to water, shade, and barn	10	_____
3. Feeding efficiency, grain, body weight, fall milk	15	_____
4. Young stock pasture, kind of pasture, body condition	5	_____
5. Availability and use of records	5	_____
	100	
D. SOUNDNESS OF FARM PROGRAM Up to 10 points may be deducted from total score, if the following conditions prevail: (a) poor use of land, labor, and machinery; (b) extravagant operation.		
Final Score:		

Mr./Ms.____________________ Your roughage program has been judged ____________________.

To improve your program, you may wish to consider the points checked below, but first call your county agent for further information.

1. Clipping after grazing
2. Smaller plots
3. More lime needed
4. More fertilizer needed
5. Super in gutter
6. Feed dry hay with lush pasture
7. Bromegrass on best land
8. Reed canarygrass on wet land
9. More Ladino clover
10. Better young stock pasture
11. More or less grain with pasture
12. Irrigation
13. Overgrazing or undergrazing
14. Alfalfa in mixtures
15. Pasture clearing
16. Mow curing
17. Sudangrass (1 acre per 8 cows)
18. Grass silage
19. Drinking water
20. Birdsfoot trefoil
21. Sign up in Soil Cons. District
22. Corn silage (1 acre per 3 cows)
23. DHI testing
24. Rye pasture (1 acre per 5 cows)
25. Trench silo

Ratings

Excellent 90–100
Good 80–90

Judges____________________
____

In Circular A303, North Dakota State University makes the following general recommendations for making quality hay and for judging hay quality.

Hay-making practices

1. To preserve carotene (vitamin A) and that good quality indicated by green color, hay must be cured as rapidly as possible and baled or stacked as soon as ready.
2. Saving the leaves is important, especially in legumes. Alfalfa leaves contain about twice as much protein and minerals as the stems. You can cut down the leaf loss by partly curing hay in the swath and then windrowing before it bleaches and dries.
3. Hay is safe to stack when the moisture content is less than 25 percent. *Remember, the stems dry out last.* If the stems are brittle enough to break when you twist a handful of hay, the hay is dry enough to stack. To prevent molding, have hay drier for baling than for loose stacking. [Baled hay should be 20% or less in moisture.]
4. When your hay is stored, take a look at it and judge the quality by using the [point system found in Table 23.2].

Table 23.2

Hay Quality Score Card

Factors	Legumes	Grasses (Points)	Grass and Legume Mix
LEAFINESS—Applies to legumes only. Top-quality hay contains 40% or more of leaves and has at least one-fifth of the leaves clinging to the stems	25	—	15
COLOR AND AROMA—Hay should be green and bright, and have a good odor	25	30	25
FOREIGN MATERIAL—Low content helps insure greater consumption and less waste of hay	15	20	20
SOFTNESS AND PLIABILITY—This is an indication of early cutting for high quality, digestibility, and palatability	15	30	20
CONDITION—Hay that is cut, cured, and stored properly will be free from dust and moldiness	20	20	20
TOTAL SCORE	100	100	100

Fig. 23.1—Artificial drying of hay helps to preserve its quality.

4. Judging Silage Quality

Making quality silage from corn, sorghum, or other forage plants is a complementary or supplementary practice, or both, of any good pasture program.

General suggestions

1. The best silage is made from crops having a moisture content of around 70 percent.
2. Before ensiling, these crops should be wilted (especially legumes) or else have chemical or grain preservatives added.

3. If properly preserved, silage will keep for long periods, so in addition to its use for seasonal feeding, it may be stored for emergency.

4. For top quality, time of harvest is just as critical for silage crops as for hay.

Table 23.3

Grass Silage Score Card (Grasses, Legumes, or Combinations of Both)

	Possible Score	Score Given
CROP QUALITY (40 points) BASED ON STAGE OF GROWTH AT CUTTING		
1. Before blossom or early heading (fine stems, high leaf content)	36-40	______
2. Early blossom	31-35	______
3. Mid-to-late bloom	21-30	______
4. Seed stage (very stemmy, coarse, low leaf content)	10-20	______
PRESERVATION (60 points) BASED ON COLOR AND ODOR		
A. COLOR (30 points)		
1. DESIRABLE—Natural forage green or slightly yellowish-green. Light to dark green depending on crop and/or additive used. Red clover may have a darker color	26-30	______
2. ACCEPTABLE—Deep, dark green or very yellowish-green or slightly brownish-green	16-25	______
3. UNDESIRABLE—Brown or black indicating excessive heating or putrefaction. Predominantly white or gray indicating excessive mold	5-15	______
B. ODOR (30 points)		
1. DESIRABLE—Clean, pleasant with no indication of putrefaction	26-30	______
2. ACCEPTABLE—Somewhat strong, yeasty, fruity or musty, slightly burnt odor, sweet	16-25	______
3. UNDESIRABLE—Strong burnt or caramelized odor indicating excessive heating. Sliminess and a putrid odor indicating improper fermentation. Very musty or moldy odor with excessive mold visible	5-15	______
TOTAL SCORE	100	______

MOISTURE CONTENT: High-moisture silage (75% or above) will contain less feed value per pound than lower-moisture silage. High moisture may indicate excessive juice loss with loss of nutrients. However, heavy nutrient loss may result from ensiling material too dry to pack well. Moisture content can be approximated by squeezing in the hand; if juice runs free, it is high moisture.

Table 23.4

Corn and Sorghum Silage Score Card

	Possible Score	Score Given
CROP QUALITY (50 points) BASED ON GRAIN CONTENT		
1. HIGH proportion of corn grain to stalks and leaves	46-50	______
2. MEDIUM proportion of corn grain to stalks and leaves	36-45	______
3. LOW proportion of corn grain to stalks and leaves	26-35	______
4. NONE (either no corn ears have developed or ears have been removed)	20-25	______
PRESERVATION (50 points) BASED ON COLOR AND ODOR		
A. COLOR (25 points)		
1. DESIRABLE—Natural green to olive green color	21-25	______
2. ACCEPTABLE—Yellowish-green to slightly brownish; if frosted, faded light yellow; slight mold spots evident	11-20	______
3. UNDESIRABLE—Deep brown or black indicating excessive heating or putrefaction. Predominantly white or gray indicating excessive mold development	5-20	______
B. ODOR (25 points)		
1. DESIRABLE—Clean, pleasant with no indication of putrefaction	21-25	______
2. ACCEPTABLE—Yeasty and fruity indicating a slightly improper fermentation; slightly burnt, sweet, caramelized, or musty odor indicating excessive air. Quite rank or sour indicating high moisture	11-20	______
3. UNDESIRABLE—Strong burnt odor indicating excessive heating. Putrid odor indicating improper fermentation. Very moldy or musty odor with excessive mold visible throughout the silage	5-10	______
TOTAL SCORE	100	______

SCORING: Excellent silage 90-100, good silage 76-89, fair silage 60-75, poor silage below 60.

CHAPTER 24

Fences for Pastures

Except on the open range, adequate fences are essential for any profitable livestock enterprise. Lifelong enemies have been made because cattle or other livestock have broken through a "line" fence and have damaged crops on another farm. Also, it's necessary to confine livestock to certain fields on an individual farm. Rotational grazing depends on a fencing system. In general, well-fed livestock are not as troublesome about breaking out of fields as hungry livestock.

Mountain States Regional Publication No. 2R, in discussing the purpose of fencing, says:

> Pasture fencing, once limited to farms and small ranches on highly productive land, is gaining wide acceptance on lands of lower productivity. Many ranchers have begun converting from herding with high cost labor to running both cattle and sheep loose in fenced pastures. Considering the large numbers of livestock run by many range operators and the vast acreages involved, converting to fenced pastures is quite an undertaking. It requires an operator to revise his fencing methods and to adapt them to range conditions.
>
> Sheep ranchers find they increase the grazing capacity of their range when they fence it into small pastures. Under fence, sheep spread over all the pasture, grazing spots herded sheep would not reach. Each ewe sets her own grazing pace and is not forced to keep up with the band, eating dust and trampled grass. As a result, ewes produce more milk and their lambs are stronger and heavier at weaning time. Sheep under fence also have cleaner wool than those traveling in herded bands.
>
> Most ewes in small pastures are "loose-lambed," that is, they are put in pastures and left alone until they finish lambing. Stockmen say this nearly eliminates "big head" and other lambing problems. Ewes left unmolested raise a greater number of twins, being less likely to desert one lamb. Ranchers who loose-lamb usually shear ewes about two weeks before lambing to reduce the occurrence of "ticky lambs." Early shearing also re-

Courtesy, Union Pacific Railroad

Fig. 24.1—Herded ewes produce less than those fenced in small pastures.

duces the chances of ewes rolling onto their backs and not being able to right themselves.

Cattle ranges have been controlled by fences for years but researchers find that cross fencing large pastures to make smaller ones can increase grazing capacity 25 percent or more through improved utilization. Compared with season-long annual grazing, a combination of deferred grazing, and rotational pastures produces the greatest increase in utilization while rotation alone produces an intermediate increase. Cross fencing and rotational grazing are essential for proper utilization of irrigated pasture. On range with suitable vegetation, operators can graze cattle and sheep in the same pastures because their eating habits differ.

Cattle ranchers find they can obtain greater bull efficiency in small pastures than in large ones. Many operators running cattle on public range use small breeding pastures to improve breeding efficiency.

The proper pasture size depends upon topography, soil, vegetation, class of livestock, and number of stock. Ranchers should consult local range management technicians regarding location of cross fences to obtain favorable pasture size and forage utilization.

Courtesy, Wichita Mountain Wildlife Refuge

Fig. 24.2—Ranges have boundary fences and cross fences to increase productivity.

Fencing pastures and rangeland has several advantages:

- eliminates need for herders
- increases grazing capacity of land as much as 25 percent
- permits grazing both cattle and sheep in the same pasture
- permits rotational and deferred grazing so pastures can be rested
- results in heavier calves and lambs in the fall
- results in cleaner wool
- permits loose-lambing and calving
- permits segregation of classes and kinds of animals
- controls straying and trespassing
- permits seasonal control of hazardous areas such as bogs or poisonous plants
- protects new seedlings until they are established

When sheep are switched from herding to fenced pastures several problems may arise. Trailing damage may occur in the pasture or along the fence lines if the pastures are too small. Where the land lacks natural protection, winter losses may be heavier under fencing unless shelter or break-away fences or gates are provided in the down-wind corner of the pasture.

Losses from predators may increase under fenced pasture management, especially at lambing time. Predators must be controlled and thefts supressed to minimize losses.

Sheep run in pastures tend to become wilder after two or three years and are harder to work than herded sheep. The operator converting from herding to pasture should anticipate the need for good working facilities. He may also need to develop additional water to use his pastures properly.

The principal deterrent to fencing as far as cattle and sheep are concerned is the initial cash outlay. Boundary fencing costs must be borne by the stockman alone or jointly with his neighbor. Those who build cross, drift, and other fences to help control and distribute livestock may be eligible for cost-share assistance through the U.S. Department of Agriculture's conservation program. Operators may obtain details of this assistance from the county agricultural stabilization and conservation committee.

Activities Which Involve Approved Practices

1. Deciding on the type of fence.
2. Deciding on the kind of fence.
3. Selecting fence posts.
4. Building the fence.

1. Deciding on the Type of Fence

The type of fence a farmer should build depends to a large extent on the following factors:

1. The type of livestock to be pastured.
2. Value of the land.
3. Climate, topography, and soils.
4. Availability of labor for construction and maintenance.

If the land is relatively valuable, the fence should occupy as little space as possible. The type of fence should be governed to some extent by the size of the animals. It would be foolish, for instance, to try to use barbed wire for confining small pigs.

2. Deciding on the Kind of Fence

There are several different kinds of fences: rail fences, board fences, stone fences, barbed wire fences, woven wire fences, electric fences, and "living" fences.

Rail fences

These picturesque zig-zag fences have just about disappeared from the agricultural scene. They require considerable space and often are damaged by heavy winds. The rail fence can, of course, be moved from one area to another without damaging the rails.

Board fences

Board fences are usually found extensively in the horse-growing area. When painted or stained to enhance color, they add to the appearance of the farmstead. Except where lumber and labor are cheap, this is an expensive type of fence to build and maintain.

Stone fences

Stone fences vary in design. They range in height from 3½ to 5 feet. Like the rail fence, they are disappearing from the farm land-

Fig. 24.3—There is a fence for every purpose. Purpose and attractiveness can be blended together.

scape although numerous stone fences are still seen in the New England area.

Barbed wire fences

There are five standard styles of barbed wire and one standard style of two-strand, twisted, barbless wire. Styles of barbed wire differ in the shape and number of points of the barb, spacing of the barbs, and size of wires.

Barbed wire is used extensively for fencing in cattle. However, in the South, barbless wire is often preferred, because there is danger of screw worm flies laying eggs in the open wounds sometimes made by barbed wire.

Barbed wire is usually sold in 80-rod spools. In purchasing the barbed wire, farmers should consider both the price per spool and the weight of the wire. The heavier the wire, the more expensive it is. Spools are clearly labeled and show weight as well as length. The general specifications for galvanized barbed wire and barbless wire are given in Table 24.1.

Table 24.1

Barbed Wire Specifications[1]

	Barbs					
Line Wire Gauge	Shape	Points	Wire Gauge	Shaping (inches)	Wraps on Line Wire	Approx. Wt./80 Rd.
12½	half round	2	14	4	1	76
12½	round	2	14	4	2	80
13½ H.T.[2]	round	2	14	4	2	64
14	round	2	16	4	2	52
12½	flat	2	12½	4	1	77
12½	round	4	14	5	2[3]	88
12½	half round	4	14	5	1	83
13½ H.T.[2]	round	4	14	5	2[3]	71
15½ H.T.[2]	round	4	16	5	2[3]	41

[1]*Source:* Mountain States Regional Publication 2R.
[2]High tensile strength wire.
[3]Wrapped around both strands; interlocked with one barb projecting between the strands.

Courtesy, USDA

Fig. 24.4—A sturdy fence keeps livestock from straying.

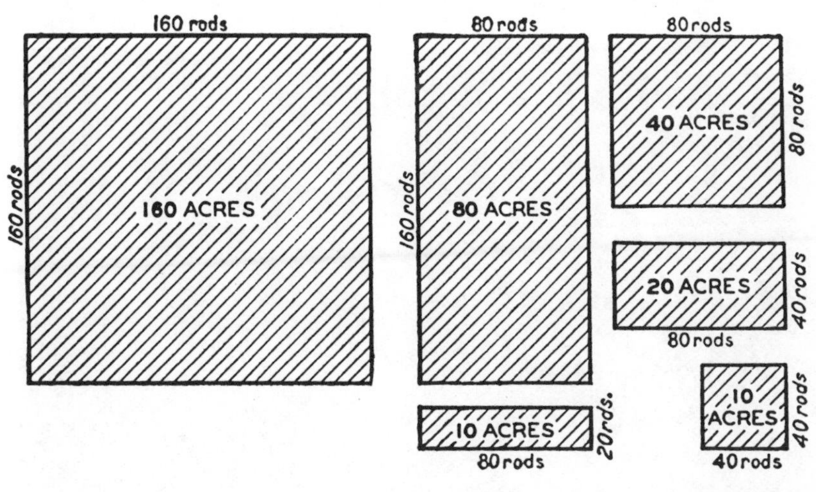

Courtesy, USDA

Fig. 24.5—The length of fencing required per acre varies with the size and shape of the field.

High-tensile wire fences

High-tensile wire fencing is increasing in popularity. Self-tightening springs, fewer line posts and reduced upkeep are apparent advantages. Smooth wire is used. This fence can be designed to control small and large animals, such as sheep, cattle, or horses.

Woven wire fences

There are eight principal standard types of woven wire fences. The most common ones are numbered 1155, 1047, 939, 832, and 726. The first two digits in the four-digit style number and the first digit in the three-digit style number designate the number of horizontal wires in the fence. The last two digits in either style number designate the height of the fence in inches. For example, Style No. 1047 has 10 horizontal wires and is 47 inches high. Woven wire fencing comes in rolls of 20 rods.

Three other types, numbered 949, 845, and 635, are the same as 1155 except that in the 949, two bottom wires are omitted; in the 845,

Courtesy, USDA

Fig. 24.6—A well-fenced, productive pasture.

Fig. 24.7—This rig speeds fence building for the smaller operator.

three bottom wires are omitted; and in the 635, five bottom wires are omitted.

The vertical stay wires are spaced 6 to 12 inches apart in standard fencing. The stay and intermediate line wires are the same size and are 11, 12, or 14½ gauge, and the top and bottom wires are 9, 10, or 11 gauge.

The standard specification numbers for woven wire fences are the same as the gauge numbers used in the fences and are given in Table 24.2

Electric fences

An electric fence consists of one or more wires carrying sufficient voltage to cause a disagreeable shock to animals that touch it. Regardless of how low a voltage, electric fences can be dangerous if improperly used. Some states have regulations regarding the design and uses of electric fences. The following safeguards, principal uses, advantages, and disadvantages are taken from U.S. Department of Agriculture Farmer's Bulletin 1832.

Table 24.2

Standard Design of Woven Field Wire[1]

Height (inches)	No. of Horizontal Wires	Spacing of Vertical Stay Wires (inches)	Gauge of Top and Bottom Wires	Gauge of Intermediate Wires	Approximate Weight per 20-Rod Roll (lbs.)	Design No.
55	11	12	9	9	342	1155-12-9
49	9	12	9	9	288	949-12-9
47	10	6	9	9	416	1047-6-9
			9	11	280	1047-6-11
			10	12½	190	1047-6-12½
		or 12	9	9	306	1047-12-9
			9	11	212	1047-12-11
			10	12½	146	1047-12-12½
45	8	12	9	9	258	845-12-9
			9	11	180	845-12-11
			10	12½	126	845-12-12½
39	9	6	9	11	246	939-6-11
			10	12½	168	939-6-12½
			11	14½	112	939-6-14½
		or 12	9	9	270	939-12-9
			9	11	188	939-12-11
			10	12½	132	939-12-12½
35	6	12	9	9	194	635-12-9
			9	11	140	635-12-11
			10	12½	100	635-12-12½
32	8	6	9	11	214	832-6-11
			10	12½	148	832-6-12½
			11	14½	98	832-6-14½
		or 12	9	9	236	832-12-9
			9	11	166	832-12-11
			10	12½	116	832-12-12½
26	7	6	9	9	266	726-6-9
			9	11	184	726-6-11
			10	12½	128	726-6-12½
			11	14½	86	726-6-14½
		or 12	9	9	202	726-12-9
			9	11	144	726-12-11
			10	12½	102	726-12-12½

[1]*Source:* Mountain States Regional Publication 2R.

1. Maximum current output must be limited. It is the quantity, or amperage, of current passing through the body that is responsible for injuries from electric shock. A properly designed controlling device will control the amperage with whatever voltage is used. On the other hand, if the amperage is not properly limited by the controlling device, even low voltages may cause dangerous shock if the subject is standing on wet ground or in contact with a grounded object. The amount of current required to cause people to "freeze" to the conductor varies with individuals, but it seems that "freezing" can be prevented by limiting current to 1⁄100 ampere—10 milliamperes.

2. Interrupted current with duration of charge not to exceed 1⁄10 second. The intervals between charges should be long enough to give an individual in contact with the fence time to free himself or herself.

3. Controlling mechanism designed so that the circuit-breaking device cannot stop in closed position.

4. Current controller sealed in a tamperproof box.

5. Housing and non-conducting parts of controller effectively grounded.

6. Only those models of fence controllers which have been approved by Underwriters' Laboratories, Industrial Commission of Wisconsin, or other such generally recognized institutions should be used.

7. Full instructions for installation and use furnished by the manufacturer of electric fences with each controller. These instructions should be followed faithfully.

8. The fence must be equipped with lightning arresters to prevent loss of property and livestock.

The principal uses of electric fences are:

1. For supplementary fencing; for temporarily enclosing pastures, feed lots, or haystacks; or for dividing off fields of corn or other crops to be hogged down.

2. For placing inside bullpens to keep bulls from damaging the heavy fences which should always be used.

3. To discourage prowlers and predatory animals.

4. For fencing marginal land which does not warrant an investment in permanent fencing, for fencing in areas where good fence posts are scarce, and for fencing in rocky land where it is hard to dig fence-post holes.

5. As a protective measure against depredation of wildlife, electric fences have been used with variable success. Four charged wires

have kept bears out of apiaries, and two wires may keep deer and antelope away from haystacks.

Advantages claimed for electric fences are:

1. Low cost. The relative cost of an electric fence and a conventional barbed wire fence will vary with the length of the fence. In long fences the cost of extra wires and posts for conventional fencing may more than equal the cost of the controller, connections, and insulators for electric fencing. The amount of electrical energy consumed by electric fences is practically negligible, but the upkeep includes the cost of keeping vegetation cut along the fence.
2. One wire required. This claim may be correct where stock of one kind and one size is to be restrained but not if the stock is of different kinds and sizes.
3. Easily moved.

The principal disadvantages of electric fences are:

1. A painful shock results from contacting the live fence wire.
2. Livestock require initial training to respect charged wires.
3. The current may fail and allow the stock to get out.
4. An electric fence cannot be depended upon to stop infuriated bulls or stampeded animals.
5. Weeds, grass, or shrubbery may ground the fence wire, and it may be necessary to cut weeds and grass several times a season.
6. Some types of electric fences need adjustment to weather conditions. Fences in dry soils require higher voltage than fences in wet soils. With some designs, the higher voltage used in fences in normally dry soils is dangerous in wet weather.
7. Some units may interfere with radio and television reception.

Wire

Just as there are several kinds of fence to control different classes of livestock, so there are different designs, styles, weights, and strengths of wire for specific jobs.

There are also coatings to protect the wire from rust and corrosion. Galvanizing is a thin coat of zinc applied to wire to delay rusting. It is measured in ounces of zinc per square foot of wire surface. The American Society for Testing and Materials has established minimum amounts of galvanizing for different wire sizes to give three classes of protection. These are shown in Tables 24.3 and 24.4.

"Living" fences

Some years ago, interest was strong in "living" fences, especially those made from multiflora rose. This beautiful plant is of the vigorous-growing bramble type and, once it reaches proper size, can

Table 24.3

Minimum Amounts of Galvanizing on Wire[1]

Wire Size (Gauge No.)	Amount of Galvanizing (Minimum Coating) (Oz. of Zinc per Sq. Ft. of Wire Surface)		
	Class I	Class II	Class III
9	0.40	0.60	0.80
11	0.30	0.50	0.80
12½	0.30	0.50	0.80
14½	0.20	0.40	0.60

[1]*Source:* Mountain States Regional Publication 2R.

Table 24.4

Approximate Protection Given Wire by Class I and Class III Galvanizing[1]

Wire Size	Years Until Rust Appears						Years After Rust Appears Until Wire Reaches Half Strength		
	Climatic Condition						Climatic Condition		
	Dry		Humid		Coastal & Industrial		Dry	Humid	Coastal & Industrial
	Class I	*Class III*	*Class I*	*Class III*	*Class I*	*Class III*			
9	15	30	8	13	3	6	50+	50+	25
11	11	30	6	13	2	6	50+	50	16
12½	11	30	6	13	2	6	50+	35	12
14½	7	23	5	10	1.5	4.5	50	20	7

[1]*Source:* Mountain States Regional Publication 2R.

very effectively confine farm animals of all sizes. The advantages claimed for this fence were:

1. Low maintenance in time and money.
2. Provides shelter for wildlife.
3. When it blooms, it adds beauty and attractiveness to the farmstead.

Here are its disadvantages:

1. It takes four to five years to reach a size that will confine farm animals. This usually means that a pasture, for instance, must be fenced with another type of fence until the multiflora rose is high enough.
2. The "living" fence takes up considerable space, often a strip 12 feet wide.
3. Plants missing or killed cause gaps in the fence. These gaps must be filled in with something else.
4. The rose thorns are painful to both humans and animals.
5. Multiflora rose has become a pest in permanent pastures, woodlands, drainage ditch banks, and fence rows.

Today, multiflora rose is considered a weed species.

3. Selecting Fence Posts

The longevity of wire and board fences depends largely on the sturdiness and durability of the posts used. Most common, of course, are wood posts. Metal posts are the second most popular, and rigid vinyl posts are increasing in use.

Wooden posts

Woods such as black locust, cedar, and osage orange have been long prized for fence posts because they are long-lasting when used for this purpose. Unfortunately, posts of these types are not available throughout the entire country.

However, modern chemical wood preservatives are effective in converting less durable woods into long-lasting fence posts. Advisability of treating wood posts, of course, depends upon cost and the availability of other kinds of posts.

Fig. 24.8—Long-lasting fence stiles are easy to build.

Of the several available preservatives, most common are coal-tar, creosote, and petroleum oil containing either phenylchlorophenol or copper naphthene and zinc chloride.

In areas where posts decay rapidly, the practice is to treat the entire post. Where climates are cool, a common practice involves treating only the butt end of the post, so that the treated area extends 6 to 8 inches above the ground line when the post is set.

Initial cost of wooden posts may be lower than the cost of metal. Also, ease of stapling the wire to wooden posts is another advantage not enjoyed by posts made of metal or rigid vinyl.

The U.S. Forest Products Laboratory has divided woods into the following classes, depending on their suitability for making fence posts:

Class 1. Woods of high durability that will probably remain in service longer than 15 years.

Hardwoods: black locust, osage orange.

Class 2. Woods of intermediate durability that will probably last from 7 to 15 years.

Hardwoods: catalpa, oak (white), red mulberry, sassafras, walnut (black).
Softwoods: baldcypress, cedar (various species), red cedar (eastern or western), redwood.

Class 3. Woods of low durability that will probably last from 3 to 7 years.

Hardwoods: ash (various species), aspen (popple), basswood, beech, birch (various species), box elder, butternut, cottonwood, elm (various species), hackberry, hickory (various species), honey locust, maple (various species), oak (red), sweet gum, sycamore, willow (various species).
Softwoods: balsam fir, Douglas fir, hemlock (eastern), larch (western), pine (various species), spruce (various species), tamarack.

Metal posts

Metal posts are made in several shapes, and there are a number of devices used for fastening the fence to the post. Although the cost of metal posts is usually higher than for wood posts, their use protects animals against lightning by grounding the current (all fences should be grounded). What's more, the metal posts are light in weight and easy to handle, and can be driven into the ground, thus reducing the labor required in building the fence. They can be used over and over again.

Rigid vinyl posts

Strength is the most limiting aspect of vinyl posts. Technological improvements may make these posts nearly equal to wood or steel. Vinyl deteriorates very slowly and does not conduct electricity. Rigid vinyl posts have been very useful for electric fencing. Electric fencing with light woven wire or chicken netting have been successful with sheep flocks.

4. Building the Fence

Details and complete coverage of fence-building techniques have not been undertaken in this book. However, the essentials as presented in Mountain State Publication R2 are reproduced in this section.

Laying out the fence line

Open, level areas make marking the line relatively simple:

1. Set a stake at each end of the proposed fence line or erect end fence assemblies.
2. Station someone at one end of the line to sight between the ends and align intermediate line stakes.
3. Set stretch braces where needed and set one or two intermediate stakes between braces. (A standard roll of barbed wire contains 80 rods of wire; this is the maximum that should be hand-stretched.)
4. Stretch a barbed wire along the ground between braces; make sure it touches the intermediate stakes.
5. Set posts along the wire at the desired spacing.

Hand construction methods have changed little over the years. About one-half of the total fencing cost is for materials and the rest is for labor. Fencing costs can be reduced by using new, improved wire, fasteners, or posts, or by altering posts or wire spacing to reduce the materials required. Operators can greatly reduce labor costs through mechanization. Anyone planning a major fence-building project should carefully consider labor-saving aspects of mechanization.

Clearing the line

Remove shrubs and trees from the fence line before building a fence. Wire nailed directly to trees is not satisfactory. Tree sap increases corrosion of the wire and staples; it is also difficult to align the fence and stretch the wire properly.

With a blade, clear a right-of-way wide enough to permit a vehicle to drive along the fence. Boundary fences may need a vehicle lane on both sides. A cleared fence line is cheaper to maintain and easier to inspect than one that is not cleared.

Corners and braces

Fence-corner construction is the critical factor in the successful performance of a wire fence. There are many designs for end, corner, and brace assemblies, but research shows one or two basic designs give maximum holding power. These will be discussed. Braces and corners are usually constructed of wood but may also be of commercial steel design. Construction methods depend upon materials used.

Steel corners and diagonal braces are usually set in concrete, but some designs have anchors on the bottom of the post and can be dirt-set the same as a wooden post.

One type of steel corner has a brace plate attached to the end of the diagonal. Drive this type into undisturbed soil with a sledge. When the brace plate is half-way into the ground, dig a narrow trough under the diagonal. Then drive the plate until the top is 2 or 3 inches below the surface.

Another commercial design uses driven steel posts and a horizontal steel brace. This brace is further strengthened by a diagonal brace of stranded wire, attached to a driven steel plate, and tightened by a turnbuckle. Use this type for end, corner, and stretch brace assem-

Fig. 24.9—Corner posts in pastures need to be well braced, neat, and strong in order to prevent sagging of the fencing.

blies. When power equipment is available, the driven steel brace can be erected in much less time than the conventional wooden brace.

Set corner posts, either steel or wooden, at least 42 inches deep. A post set 42 inches has double the holding power of one set only 30 inches.

Horizontally braced corners with a diagonal brace wire are the simplest and most effective corner and brace assemblies.

Wooden brace posts should be at least 5 to 6 inches in diameter and set 42 inches deep. Use a wooden horizontal brace at least 4 inches in diameter or a 2-inch steel pipe. The posts should be 6 to 8 feet apart and perfectly in line with the fence. Use a double loop (four strands) of No. 9 wire or two strands of twisted 12½ gauge barbless wire to make the diagonal wire brace.

To construct:

1. Set corner or end post at least 42 inches deep, tamping after each shovel of dirt.
2. Cut horizontal brace. Anchor the brace to the corner post with a steel dowel to prevent the brace from kicking out if the assembly is twisted. For a slightly less effective assembly, notch the post to give a flat bearing surface and nail the horizontal brace securely.
3. Set the second post at the end of the horizontal brace and notch or drill for dowel.
4. Attach the brace wire as shown in Fig. 24.9 and twist tight with a stick or rod until the whole assembly is rigid. Leave one end of the stick long enough to anchor behind the horizontal brace to prevent the wire unwinding.

Setting line posts

In average soil conditions, drive steel line posts so the anchor is about 6 inches below the surface. Use an ordinary sledge or, better yet, a post drive—a pipe with a weight on top. For rapid (about 30 posts per hour), easy fence building drive posts with a power post driver. Select posts of proper length; the top wire should be near the top of the post.

Wooden posts, usually set in a dug hole, can also be driven with a power post driver into any soil you can dig with a shovel. One person can drive about 15 wooden posts an hour with a power driver. The person would average six posts an hour with a power digger and hand tamping. A driven post is firmer than a hand-tamped post, and in sandy soil there is less danger of the wind whipping soil from around the post. To be power driven, posts must be relatively straight. Under gravelly soil conditions, posts more than 4 inches in diameter should be sharpened.

Set wooden posts 24 to 30 inches deep in the center of the hole so they can be tamped all around. Posts set against the side of a hole will always be loose because soil cannot be properly tamped.

Placing the wire

After setting the posts, unroll the wire on the side of the posts which will have the most livestock pressure. Place woven wire with the bottom of the roll (close mesh) next to the posts. Start about 2 feet ahead of the anchor post and unroll 30 to 40 feet of wire; stop and attach it to the anchor post. Unroll the wire to the next brace assembly.

To unroll barbed wire, place a steel bar through the center of the

Fig. 24.10—Horse pastures need special fencing precautions. Cables and rods are often used to reduce the possibility of cuts and bruises.

roll and secure the end of the wire to the anchor post. If using more than one strand, place several rolls on the same bar with a metal disc between the rolls to prevent fouling. Put the bar across the bed of a pickup truck and anchor it to stakes in the stock rack mounting holes. The wire unrolls as you drive the pickup along the fence line.

On small jobs, unroll single rolls by hand with a person at each end of the bar. Use a bar with a slight bow to help keep the roll in the center of the bar.

Splicing the wire

The "wrap" or "western union" type splice is strong, will not damage or break the wire, and is easy to make, either by hand or with a simple splicing tool.

When splicing woven wire, leave about 6 inches of wire beyond the vertical stay; place the vertical stays together and wrap the free ends of wire behind the vertical stays. Some prefer to overlap the wire one whole section to provide a better handhold for wrapping the splice.

Stretching the wire

More wire is damaged by overstretching than by any other cause. Be extremely cautious when using a tractor to stretch wire. This practice may not only damage the wire but also will endanger anyone standing near the fence if the wire breaks. For protection, always stand on the side of the post opposite the wire.

Woven Wire—Attach a clamp bar to the wire opposite the brace post. Attach stretchers to the clamp, top, and bottom, and to the anchor post. If using only one stretcher, attach it to the clamp so there are as many wires above the point of attachment as below. Handstretch the wire between the clamp and the anchor post for final attachment. Some prefer to stretch the wire past the anchor post by attaching the stretching blocks to a truck, tractor, or dummy post.

Slowly tighten the wire so the tension is evenly distributed along its length. A second person can help by shaking the fence loose from posts or obstructions while it is being stretched. Propping the fence against the posts with temporary stakes will prevent the wire from crowding down on the ground, especially at ridges. Place the stakes under the top wire and lean them slightly out and away from the direction of pull.

In warm weather, woven wire is properly stretched when it is springy to the touch and the tension curves begin to straighten. When erecting fence in temperatures below 40 degrees remove not more than half the tension curve. This will allow the fence to expand and contract with temperature changes.

Barbed Wire—There is no good way to tell when barbed wire is tight enough. Stretch it until it is springy to the touch, or support the wire by wire loops or spikes at the top of the posts at intervals of 8 to 10 posts and stretch the wire until it sags about 12 inches between supports. Stand clear of barbed wire while stretching it to avoid serious injury should the wire break.

Fastening the wire

Attaching the wire to wooden posts is one of the most important steps in fence building.

On line posts, slope the staples against the pull of the wire. On level ground and over knolls, slope the staples slightly downward. In draws where the wire tends to lift, slope the staples upward.

Make the staples more secure by angling them slightly from parallel to the grain of the wood. This will reduce splitting, especially in small, treated posts. Using L-shaped staples with deformed shanks will improve holding power in treated posts. If the deformed staples are not available use 1½- to 2-inch galvanized staples with treated posts. Galvanizing reduces rusting and increases holding power.

Do not drive staples tight. The wire should always be free to move through the staple. Tight stapling damages the wire at the point of contact and also limits seasonal contraction and expansion. Except at corners, never place the staple over or adjacent to the barb or the stay on woven wire.

When stapling woven wire, staple the top and bottom wires on each post first. It is not necessary to staple all intermediate wires on most woven wire fences.

When fastening wire to the posts, start midway between the corners or ends and staple toward each end. This will give a uniformly tight fence.

Gates and passageways

Gates in range fences are usually made of wire to match adjoining fence. Livestock gates should have at least one strand of barbed wire on top. The height of the wire on the gate is about the same as the line fence. Gate width is determined by the number and kind of livestock and by the width of vehicles to pass through the gate. A 12-foot gate is sufficiently wide for trucks and cars and will handle moderate numbers of livestock. If farm machinery will pass through the gate, it should be 14 to 16 feet wide. This is also wide enough for large numbers of stock.

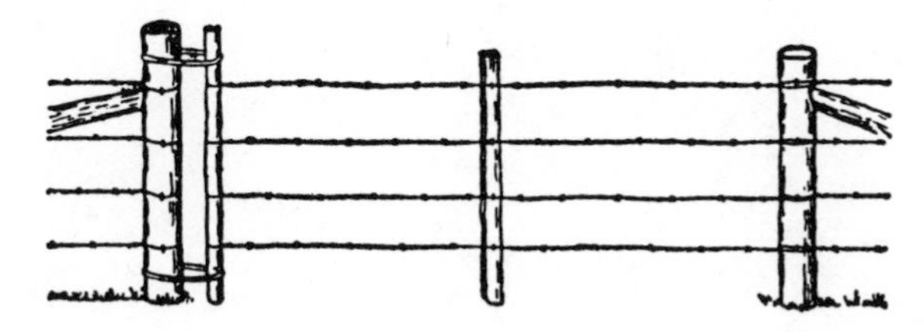

Courtesy, USDA

Fig. 24.11—Follow this illustration to make a good, inexpensive gate.

Make wire gates by attaching the wire to 3-inch-diameter poles or stakes and attaching the gate stakes to the fence anchor poles with loops of wire. For a sheep-tight gate, place the gate stakes 4 to 6 inches from the fence post. If the gate is more than 12 feet wide, add an intermediate stake to keep the wires from sagging. If the gate sags when closed by hand, use a gate-closer.

APPENDIXES

APPENDIX A

Estimating Grain, Hay, and Silage[1]

GRAIN: A bushel of grain contains 1¼ cubic feet. Multiply the length of the bin times the width times the depth (all in feet), and divide by 1¼ (or multiply cubic feet in bin times 4 and divide product by 5 to obtain bushels).

EAR CORN: Multiply the length of the crib times the width times the depth (all in feet), and divide by 2½ (or for bushels of ear corn multiply cubic feet in bin times 2 and divide product by 5).

HAY: Multiply the length of the mow times the width times the height, and divide by 400 to 500 depending upon the kind of hay and how long it has been in the mow. The bottom is much heavier than the top. In the bottom 2 to 3 feet, 400 to 450 cubic feet may weigh a ton.

SILAGE IN ROUND SILOS

Depth of Silage One Month After Filling	Inside Diameter of Silo in Feet					
	10	12	14	16	18	20
(feet)	*(tons)*					
8	11	16	21	28	35	43
9	12	18	24	32	40	49
10	14	20	27	35	44	55
11	16	22	30	39	49	61
12	17	24	33	43	54	67
13	19	26	36	47	60	74
14	20	29	39	51	65	80
15	21	31	42	55	70	86

SILAGE REMAINING IN SILOS AT TIME OF INVENTORY

Depth of Silage Left	If Silage Was Less Than 20 Feet Before Any Was Fed						20-30 Feet						If Silage Was Over 30 Feet Before Any Was Fed					
	10	12	14	16	18	20	10	12	14	16	18	20	10	12	14	16	18	20
(feet)	*(tons)*																	
1	1	2	3	4	5	6	1	2	4	4	5	6	2	2	4	5	6	7
2	2	4	6	8	10	12	2	4	7	8	10	12	4	5	7	9	10	13
3	4	7	9	12	16	19	4	7	10	12	16	19	6	8	10	13	16	19
4	5	9	12	16	21	25	6	9	13	16	22	25	8	10	14	18	22	26
5	7	11	15	20	26	31	8	12	16	21	27	32	9	12	17	22	27	32
6	9	13	18	23	30	37	9	14	20	25	32	38	10	14	20	26	32	39
7	10	15	21	27	35	43	11	16	23	29	37	45	12	17	24	31	38	46
8	12	17	24	31	40	49	12	18	26	33	42	54	14	19	27	35	43	52

[1]*Source: New Holland Farm Management Book,* New Holland Division of Sperry Rand.

APPENDIX B

Forage, Seed, and Plant Characteristics and Seeding Table

Temperature type: C = cool-weather growth; W = warm-weather growth.
Growth habit; A = annual; WA = winter annual; B = biennial; P = perennial; P(A) = perennial but grown as an annual.
Photoperiodic reaction: L = long day; S = short day; N = day neutral or indeterminate; I = intermediate.

Crop	Scientific Name	Seeding Rate per Acre (close drills)	Seeds per Pound (thousands)	Weight per Bushel (lbs.)	Germination Time (days)	Temperature Type	Growth Habit	Photoperiodic Reaction
Alfalfa	*Medicago sativa L.*		220	60	7	C	P	L
Humid areas		10-20						
Irrigation		10-15						
Bahiagrass	*Paspalum notatum*	10-12	150		21	C	A; WA	L
Bentgrass								
Colonial	*Agrostis tenuis*	40-60	8,000	20-40		C	P	
Creeping	*Agrostis palustris*	40-60	7,700		28	C	P	L
Bermudagrass	*Cynodon dactylon L.*	6-8	1,800	40(14)	21	W	P	
Big trefoil	*Lotus pendunculatus*	4-6	1,000	60	7	C	P	
Birdsfoot trefoil	*Lotus corniculatus L.*	8-12	375	60	7	C	P	
Black medic	*Medicago lupulina L.*	10-15	300	60 (hulled)	7	C	A	
Bluegrass								
Annual	*Poa annua L.*	15-25	1,200		21	C	A	
Canada	*Poa compressa L.*	15-25	2,500	14	28	C	P	L
Kentucky	*Poa pratensis L.*	15-25	2,200	14	28	C	P	N
Texas	*Poa arachnifera*	6-10	1,100		28	C	P	
Blue panicgrass	*Panicium antidotale*	2-6	610		28		A	
Bluestem								
Big	*Andropogan gerardi*	15-20	150		28	W	P	S
Little	*Andropogan scoparius*	12-20	260		28	W	P	
Bromegrass								
Smooth	*Bromus inermis*	15-20	137	14	14	C	P	L
Mountain	*Bromus marginatus*	10-20	64			C	P	

(Continued)

Appendix B (Continued)

Crop	Scientific Name	Seeding Rate per Acre (close drills)	Seeds per Pound (thousands)	Weight per Bushel (lbs.)	Germination Time (days)	Temperature Type	Growth Habit	Photoperiodic Reaction
Buckwheat (common)	*Fagopyrum esculentem*	36-60	20	48	6	W	A	L; N
Buffalograss (caryopses)	*Buchloe dactyloides*	5	330					
Bur clover								
California (out of bur)	*Medicago hispida*	15-20	209	50	14	C	WA	
Canarygrass	*Phalaris canariensis L.*	20-25	68	50			A	
Carpetgrass	*Axonopus compressus*	8-12	1,350	18-36	21	W	A	
Clover								
Alsike (alone)	*Trifolium hybridum L.*	6-8	680	60	7	C	P	L
Crimson		15-25	150	60	7	C	WA	L
(unhulled)	*Trifolium incarnatum L.*	40-60						
Hop	*Trifolium agararium L.*	8-12	830	60			WA	L
Ladino	*Trifolium repens L.*	5-7	860	60	10	C	WA	L
Large hop	*Trifolium campestre*	3-4	2,500	60	14	C	WA	L
Red	*Trifolium pratense L.*	8-12	260	60	7	C	WA	L
Sub (subterranean)	*Trifolium subterraneum*	20-25	55	60	14	C	WA	
White	*Trifolium repens L.*	5-7	700	60	10	C	P	L
Cowpea	*Vigna sinensis*	75-120	2-6	60	8	W	A	S
Dallisgrass	*Paspalum dilatatum*	8-25	340	12-15	21	W	P	
Dropseed, sand	*Sporobolus cryptandrus*	3-5	5,450		42	W	P	
Fescue								
Meadow	*Festuca elatior L.*	10-25	230	14-24	14	C	P	
Sheep's	*Festuca ovina L.*	25		10-30	21	C	P	
Field pea								
(large seeded)	*Pisum arvense L.*	120-180	4	60	8	C	A	L
(small seeded)	*Pisum arvense L.*	90-120						
Gamagrass	*Tripsacum dactyloides L.*	Veg.	7			W	P	S

(Continued)

Appendix B (Continued)

Crop	Scientific Name	Seeding Rate per Acre (close drills)	Seeds per Pound (thousands)	Weight per Bushel (lbs.)	Germination Time (days)	Temperature Type	Growth Habit	Photoperiodic Reaction
Grama								
Black	*Bouteloua eripoda*	7-9	560			W	P	
Blue	*Bouteloua gracilis*	8-12	900		28	W	P	N; S
Hairy	*Bouteloua hirsuta*	10-15	980					
Side oats	*Bouteloua curtipendula*	15-20	200		28	W	P	
Indiangrass	*Sorghastrum nutans L.*		170		21	W	P	I
Johnsongrass	*Sorghum halepense L.*	20-30	130	28	35	W	P	S
Kudzu	*Pueraria thunbergiana*	Veg.	37	54	14	W	P	
Lespedeza								
Sericea	*Lespedeza cuneata*	30-40		35	28	W	P	
Common	*Lespedeza striata*	25-30	343	25	14	W	A	L
Kobe	*Lespedeza striata*	30-35	185	30	14	W	A	L
Korean	*Lespedeza stipulaceae*	20-25	240	45	14	W	A	P
Lovegrass								
Sand	*Eragrostis trichoides*	1-2	1,800			W	P	
Weeping	*Eragrostis curvula*	¼	1,500		14	W	P	
Lupine								
Blue	*Lupinus angustifolius L.*	70-90	3	60	10	C	A	N; S
Meadow foxtail	*Alopercurus pratensis L.*	15-25	540	6-12	14	C	P	
Millet								
Foxtail	*Setaria italica L.*	20-30	220	50	10	W	A	S
Pearl (Cattail)	*Pennisetum typhoides*	16-20	85		7	W	A	S
Proso	*Panicum miliaceum L.*	15-35	80	56	7	W	A	S
Napiergrass	*Pennisetum purpureum*	Veg.	1,402		10	C	P	
Oatgrass, tall meadow	*Arrhenatherum elatius L.*	30-40	150	11-14	14	C	P	
Orchardgrass	*Dactylis glomerata L.*	20-25	590	14	18	C	P	N; L

(Continued)

Appendix B (Continued)

Crop	Scientific Name	Seeding Rate per Acre (close drills)	Seeds per Pound (thousands)	Weight per Bushel (lbs.)	Germination Time (days)	Temperature Type	Growth Habit	Photoperiodic Reaction
Rape								
Oilseed (summer)	*Brassicus napus* subsp. *annua*	3-4	160		7	C	A	L
Winter (broadcast)	*Brassica napus* subsp. *biennis*	4-6						
Redtop	*Agrostis alba L.*	10-12	5,100	14	10	C	P	
Reed canarygrass	*Phalaris arundinacea L.*	8-12	550	44-48	21	C	P	L
Rescuegrass	*Bromus catharticus*	25-30	1,700	8-12	35	W		
Rhodesgrass	*Chloris gayana*	10-12	1,700	8-12	14	W	P	
Rye	*Secale cereale L.*	28-112	18	56	7	C	A; WA	L
Ryegrass								
Perennial	*Lolium perenne L.*	25-30	330	24	14	C	P	L
Sand dropseed	*Sporobolus cryptandrus*	½	3-5			W	P	
Sorghum	*Sorghum bicolor*				10	W	A	S
Kafir	*Sorghum bicolor*	15-45	20	56	10	W		
Milo	*Sorghum bicolor*		15	56	10	W		
Sorgo	*Sorghum bicolor*	15-75	28	50	10	W		
Sudangrass	*Sorghum bicolor*	20-35	55	40	10	W	A	S
Sweet clover								
White	*Melilotus alba Med.* (hulled)	12-15	250	60	7	C	B; A	
Yellow	*Melilotus officinalis* (unhulled)	30-45	250	60	7	C	B	
Switchgrass	*Panicum virgatum L.*		370		28	W	P	S
Timothy (alone, spring)	*Phleum pratense L.*	8-12	1,230	45	10	C	P	L
(alone, fall)		3-4						
(with clover)		4-6						
Vetch								
Common (alone)	*Vicia sativa L.*	40-80	7	60	10	C	A; WA	L
Hairy (alone)	*Vicia villosa*	20-40	21	60	14	C	WA; B	L
Wheatgrass								
Crested (standard)	*Agropyron desertorum*	12-20	190	20-24	14	C	P	
Slender	*Agropyron trachycaulum*	12-20	150		14	C	P	
Western	*Agropyron smithii*	12-20	110		35	C	P	L

APPENDIX C

Glossary

A horizon—The surface and subsurface soil which contains most of the organic matter.
Acid soil—A soil with a pH of less than 7.0 (usually less than 6.6).
Aerial roots—Roots arising from the stem above the ground.
Aggregate—A mass or cluster of soil particles.
Alkali soil—A soil containing alkali salts sufficient to harm crop plants.
Annual—A plant which completes its life cycle from seed in one year.
Arid climate—Usually with a precipitation of less than 10 inches.
Awn—The bristle extending from the tip of the lemma.
Axis—The main stem of a flower or panicle.
B horizon—The subsoil layer of a soil in which leached material is deposited.
Basin listing—Basins in lister furrows formed by small dams.
Biennial—A plant germinating one season and producing seed the next.
Blade—The part of the leaf surface above the sheath in grasses.
Branch—A lateral stem.
Broadcast—Random sowing of seed on land by hand or machine.
Bud—A rudimentary leaf, stem, or branch.
C horizon—The layer of weathered parent material below the solum.
Calcareous—A soil containing sufficient calcium or magnesium carbonate to give effervescence.
Capillary—A very small tube (of plant or soil).
Carotene—Yellow pigment in green leaves (the precursor of vitamin A).
Caryopsis—The grain of grasses.
Cereal—Edible seed of cultivated grass.
Chlorophyll—Green matter of plants which takes part in the photosynthetic process.
Ciliate—Hair fringe on the margin.
Clay—Small mineral soil particles.
Climate—The overall characteristic weather of any region.
Contour furrows—Channels at right angles to the slope.
Cotyledons—First leaves of a plant as found in the embryo.
Culm—The jointed stem of grasses.
Crown—The base of a stem where roots arise.
Dicotyledons—Plants producing two cotyledons in each fruit.
Drill—An implement for sowing seeds in rows.
Embryo—Rudimentary plantlet.
Endosperm—Starchy interior of grain.
Erosion—Wearing away of the land surface.
Fertility (soil)—Ability of soil to supply proper nutrients in proper proportions.
Fibrous roots—Threadlike roots of grasses.
Float—A land leveler.

Fungicide—Chemical substance used to destroy fungi.
Glumes—Pair of bracts at the base of a spikelet.
Hardpan—A hardened or cemented soil horizon, often a compacted layer of clay.
Hay—Herbage cut, dried, and cured for forage.
Hormone—A chemical growth regulator.
Hull—Structure enclosing a seed.
Humid climate—Precipitation of 30 to 40 inches yearly.
Hybrid—Offspring of two unlike parent plants.
Insecticide—Chemical used to kill insects.
Joint—A node of a stem (culm).
Kernel—A matured body of an ovule.
Leach—Removal of materials in solution.
Lemma—Outer bract of a grass spikelet.
Ligule—Membranous structure at the top of the sheath of grasses.
Limestone—Ground calcium carbonate.
Loam—A soil composed of sand, silt, and clay in such proportions that the characteristics of no one separate predominates.
Meadow—An area of fine stemmed forage plants.
Monocotyledon—A plant with one cotyledon as in grasses.
Muck—Well decomposed organic soil material, high in mineral content.
Neutral (soil)—Soil neither acid nor alkaline.
Nitrification—Formulation of nitrates from ammonia by bacteria.
Nitrogen fixation—Conversion of atmospheric nitrogen to nitrogen compounds by bacteria.
Node—Joint of a culm where a leaf is attached.
Nutrient (plant)—Chemical element essential to plant growth.
Palea—Inner, upper bract of a grass floret.
Pasture—An area covered by forage plants and grazed by animals.
  Native pasture—A forage area covered by native plants.
  Pasture succession—A series of crops grazed successively.
  Rotation pasture—An area foraged for a few seasons and then put into cultivated crops, also pasture grazed a few days then rested for several days before regrazing within the same season.
  Tame pasture—Grazing of domesticated plants.
  Temporary pasture—An area grazed during a short period only, usually annual pasture species.
Peat—Slightly decomposed organic material accumulated in a wet area.
Perennial—A plant living more than two years from original growth.
Petiole—The stalk of a leaf.
Photoperiodism—Response of plants to different day length or light periods.
Pistil—The pollen-receiving structure of a flower.
Pollination—Transfer of pollen from stamen to pistil.
Productivity (soil)—The capability of a soil to produce a given plant or sequence of plants under a management system.

Profile (soil)—A vertical section of a soil through all its horizons.
Protoplasm—The contents of a living cell.
Race—Plants having certain common characteristics due to common ancestry.
Radicle—Root of an embryo.
Reaction (soil)—Degree of acidity or alkalinity of a soil.
Rootstock—Rhizome.
Runner—A creeping branch or stolon.
Saline soil—A soil with an excess of soluble salts.
Sand—A small mineral fragment of soil larger than silt.
Seedling—Early stage of a plant grown from a seed.
Sessile—A leaf without a petiole.
Shoot—A stem with its attached members.
Silage—Forage in a succulent condition preserved by partial fermentation in a fairly air-tight container (silo).
Silt—A soil particle intermediate in size between clay and sand.
Solum—The A and B horizons of soil.
Stalk—A stem.
Stipule—Undeveloped pair of leaves at base of legume leaf petiole.
Stolon—A creeping stem above ground that produces roots and shoots.
Stool—Aggregate of a stem and its attached tillers.
Strain—A group of plants derived from a species.
Strip cropping—Growing crops in long narrow bands across a slope.
Stubble—Basal portions of stems left standing after cutting.
Subhumid climate—A precipitation of 20 to 30 inches.
Subsoil—The B horizon of a soil.
Succulent—Juicy.
Sucker—A tiller of a plant.
Symbiotic nitrogen fixation—Fixation of atmospheric nitrogen by bacteria infecting the roots of legumes.
Terminal—Upper extremity of bud, flower, or leaf.
Terrace—An earthen channel and embankment across a slope.
Tiller—An erect shoot arising from the crown of a grass plant not arising from the seed directly.
Tilth—General condition of a soil with respect to desirability for planting a crop.
Topsoil—Surface or A horizon of soil.
Transpiration—Evaporation of moisture through the leaves.
Turf—Upper portion of soil filled with roots and stems of low, living grasses, as in lawns.
Variety—A group of individuals within a species differing from the rest of the species.
Weed—A plant in its location more harmful than beneficial.
Windrow—Herbage raked into a row after cutting.
Winter annual—A fall-germinating plant that blooms the following spring or summer.

INDEX

Index

E

F

G

H

S

T

V

W

Y